Artificial Intelligence and Robotics
Ten Short Lessons

Also in the **POCKET EINSTEIN** *series*

Space Travel: Ten Short Lessons

ARTIFICIAL INTELLIGENCE AND ROBOTICS

Ten Short Lessons

Peter J. Bentley

Johns Hopkins University Press
Baltimore

First published in Great Britain in 2020
by Michael O'Mara Books Limited
9 Lion Yard
Tremadoc Road
London SW4 7NQ

Printed in the United States of America on acid-free paper

2 4 6 8 9 7 5 3 1

Johns Hopkins University Press
2715 North Charles Street
Baltimore, MD 21218-4363
www.press.jhu.edu

Library of Congress Control Number: 2020934701

ISBN 978-1-4214-3972-3 (paperback : acid-free paper)
ISBN 978-1-4214-3973-0 (ebook)

Designed and typeset by Ed Pickford
Illustrations by David Woodroffe

Special discounts are available for bulk purchases of this book. For more information, please contact Special Sales at specialsales@press.jhu.edu.

Johns Hopkins University Press uses environmentally friendly book materials, including recycled text paper that is composed of at least 30 percent post-consumer waste, whenever possible.

CONTENTS

INTRODUCTION

I grew up in the 1970s when there was no internet, no World Wide Web, and the very first affordable home computers were just emerging. In those days you had to be a serious geek like me to be into computers. Yes, I was that kid – shy at school, but prolific at home, building bizarre robots, programming early computers, writing simple computer games. I would lust after the latest computer as another child might desire a Lamborghini. So amazing, but so unaffordable! Artificial intelligence – getting my computers to think, to simulate biological behaviour and control my robots – was my childhood passion. But to the people around me, it was a passion that seemed as obscure as stamp collecting might today.

And then things changed. Quite dramatically. Today we live in a science-fiction story come true. Computers rule the world. Data floods from everything we do. Robots are building our products in factories. Our homes are computerized and we can talk to these digital home assistants, receiving detailed and coherent replies. Behind the scenes, artificial intelligence is making everything work. My obscure childhood passion is now not only mainstream, it's considered one of the most important kinds of technology being created today.

You cannot live in the modern world without interacting with, or being impacted by, AI and robots. Every time you make a purchase, AIs are handling your money, checking for fraud, using your data to understand you better, and recommending new products to you. Every time you drive a car, AIs are helping the car to proceed safely, they're watching you from road cameras and automatically changing speed limits, they're detecting your licence plate and monitoring your movements. Every time you post something on social media, AIs may trawl through the text to understand your sentiment on specific topics. As you browse the internet and read news articles or blogs, AIs monitor your activity and try to please you by feeding you more of the content you prefer. Every time you take a photo, AIs adjust the camera settings and ensure the best possible picture is taken – and then can identify everyone in the picture for you. Face recognition, speech understanding, automatic bots that answer your questions online or by phone – all performed by AIs. Inside your home you have smart TVs, computerized fridges, washing machines, central heating, air-conditioning systems – all AI robotic devices. The world economy is managed by AIs, financial trading is performed by AIs, and decisions about whether you should or should not be accepted for financial products are made by AIs. Your future anti-viral and antibacterial drugs are being designed by AI. Your services of water, electricity, gas, mobile phone and internet connections are all adjusted by smart AI algorithms that try to optimize supply while minimizing waste. You interact with a thousand AIs a day and you are blissfully unaware of them all.

In this book I'll explain a little bit about how this has happened, how it works and what it means. This is a pocket guide, so I'm going to be brief. I won't bore you with detailed technical descriptions, I will not explain every single AI technique, and I will not tell you about every AI pioneer. That would take a thousand books of this size, with more books needed every day (progress is fast!). Instead I'll take you on a short journey through this strange world of computers, robots and building brains. I'll try to point out some interesting sights along the way, and explain some of the fundamental ideas behind artificial intelligence and robotics. This journey may sometimes be a rollercoaster, for AI has its ups and downs. It has lived a surprisingly long life already, and suffered the pains of disappointment as well as the excitement of success. It is being created to change our world for the better, yet in some cases it is responsible for causing fundamental problems. Buckle up, and enjoy the ride!

Peter J. Bentley

01 A JOURNEY OF A THOUSAND MILES BEGINS WITH A SINGLE STEP

'I confidently predict that in the next ten or fifteen years something will emerge from the laboratory that is not dissimilar to the robot of science-fiction fame.'

CLAUDE SHANNON (1961)

Classical stone architecture and statues surround you. You walk through the cobbled streets of the pretty Greek island, admiring the view. The hot sun is now low in the sky, leaving a pleasant evening for a stroll around the town. The hustle and bustle of everyday life has faded away as the market stalls of fruit and fish are closed. There is just the sound of your own footsteps echoing from the ornate buildings. An unexpected movement catches your eye from the corner of the street. Yet there is nobody there. You look harder.

The stone statue – it moved! You nervously walk over for a closer look. Its chest appears to rise and fall as though it breathes. As you watch, its head turns left, then right. You realize that it's not the only one. All the statues on the streets around you seem to display some movement. They move their feet as if adjusting position, they move arms as though having some silent stone discussion. Are they slowly coming to life as night falls? Looking closely, you realize they all seem to have hidden mechanisms, cogs and wheels whirring. You're on an island of stone robots.

Ancient robots

This was the Greek island of Rhodes, as you may have found it 2,400 years ago, even before their giant statue, Colossus of Rhodes, was constructed. It was a remarkable island famed for its mechanical inventions, including life-sized automata made from marble. An Ancient Greek poet named Pindar visited Rhodes and wrote about his experience in a poem:

The animated figures stand
Adorning every public street
And seem to breathe in stone, or
move their marble feet.

It may seem inconceivable that before the Roman Empire in 400 BCE there was such ancient robotic technology. But many ancient examples are well documented. Powered by water or weights, there were mechanical lions that roared, metal birds that sang, and even mechanical people who played together in a band. King Solomon, who reigned from 970 to 931 BCE, was said to have had a golden lion that raised a foot to help him to his throne, and a mechanical eagle that placed his crown upon his head. Ancient Chinese texts tell the story of a mechanical man presented to King Mu of Zhou (1023–957 BCE) by the 'artificer' Yan Shi. Archytas, founder of mathematical mechanics, philosopher and pal of Plato, who lived from 428 to 347 BCE, made a mechanical dove – a flying, steam-powered wooden robot bird. Hero of Alexandria (10–70 CE) wrote an entire book about his automaton inventions, and how hydraulics, pneumatics and mechanics could be used. Hero even created a programmable puppet show that used carefully measured lengths of thread that were pulled by a weight to trigger different events in his choreographed mechanical play.

This fascination with building mechanical life continued unabated through the Middle Ages. Countless inventors created mechanical marvels designed to entertain. By the eighteenth century this was taken to a new level by the inventors of automated factory machines, enabling the Industrial Revolution. Laborious jobs such as weaving that had always required skilled human craftspeople were suddenly replaced by astonishing steam-powered machines that could create finer cloth faster than ever before. As one set of jobs were lost, whole new industries were created as our massive machines needed constant care and maintenance.

The decades rolled past and our expertise in building machines increased. Trains, automobiles, aeroplanes and sophisticated factories became commonplace. With an increasing reliance on automatic machines, the allure of robots and their similarity to living creatures only intensified, entering literature and movies. It is perhaps no coincidence that two of the very earliest science-fiction movies, *Metropolis* (1927) and *Frankenstein* (1931), tell the story of crazed inventors creating life.

By the twentieth century scientists were trying to understand life itself through making analogies. Perhaps, they thought, if we could make something that moved and thought like a living creature then we would learn the secrets behind life – understanding by making. This was the start of artificial intelligence (AI) and robotics as we know them today.

The birth of AI and robotics

One of the very earliest examples of an autonomous robot designed to help us understand living systems was built in the late 1940s by neurologist Grey Walter in Bristol, UK. Since they looked a little like electric tortoises, he named them Elmer and Elsie (ELectro MEchanical Robots, Light

WILLIAM GREY WALTER (1910–77)

Grey Walter pioneered robots with a mind of their own. His mechanical tortoises sensed their environments, moving towards light and away from anything they might bump into. They could even find their way back to a charging station when their battery was low. Walter was a pioneer in technologies such as the electroencephalograph, or EEG machine, for studying the human brain. He claimed these simple robots had the equivalent of two neurons and that by adding more cells they would gain more complex behaviour – something he tried by making a more complex version called Cora (Conditional Reflex Analogue) where he trained the robot to respond to a police whistle in much the same way that Pavlov conditioned dogs to salivate at the sound of a bell. The Cora robot initially made no response to the whistle, but if the whistle was blown when an electric torch was flashed, it soon learned to associate the two stimuli, and responded to the whistle on its own as though it was seeing light.

Sensitive). Grey Walter's robots were unique because they followed no specific program.

At around the same time that he constructed his experimental robots, Walter was a member of a very exclusive group of young scientists in the UK known as the Ratio Club. These neurobiologists, engineers, mathematicians and physicists would meet regularly to listen to an invited speaker and then discuss their views on cybernetics, or the science of communications and automatic control systems in both machines and living things. It was one of the very first AI and robot clubs. Most of the members went on to become eminent scientists in their fields. One of the enthusiastic mathematicians was called Alan Turing.

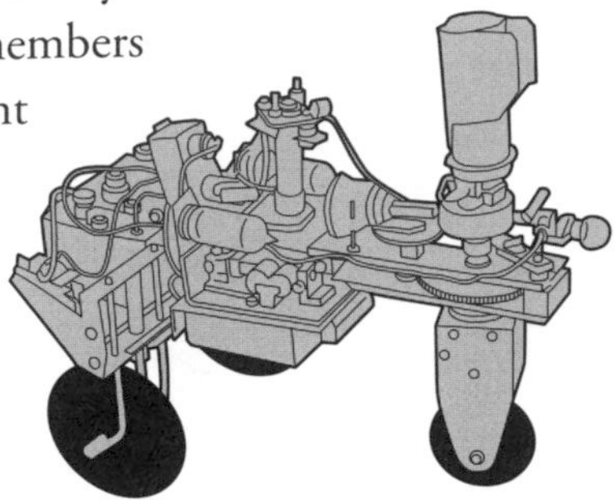

By 1950, Turing had already contributed hugely to the embryonic field of computers. His early work had provided fundamental mathematical proofs, for example that it would not be possible for any computer to predict if it might stop calculating for any given program, or in other words, some problems are not computable. He helped design the very first programmable computers, and his secret work at Bletchley Park helped decode encrypted messages during the Second World

War. Like many computing pioneers, Turing also had a fascination for intelligence. What was it? How could you make an artificial intelligence? And if you ever somehow made a computer that could think in the same way that living creatures think, how would you know? Turing decided that we needed a method for measuring whether a machine could think. He called it the 'Imitation Game', but his test became known as the Turing test.

THE TURING TEST

An interrogator can communicate with two people – each in a separate room – by typing text. He can ask any questions he likes: 'Please write me a poem on the subject of the Forth Bridge'. Or, 'What is 34,957 added to 70,764?' The two people then type their responses. After a while, the interrogator is informed that one of the two people is actually a computer. If he cannot distinguish the computer from the real person, we can then say that the computer has passed the test.

The Turing test became an important measure for AI, but it also drew much criticism. While it may provide some idea of the ability of the AI to reply to written sentences in a seemingly thoughtful manner, it does not measure many other forms of AI, such as prediction and optimization, or applications such as robot control or computer vision.

Turing was not the only pioneer of computers to think about AI. Nearly all of them did. In the US, John von Neumann, a mathematical genius who helped describe how to build the first programmable computers in 1945, worked with Turing on intelligent computers. Von Neumann's last project was on self-replicating machines, an idea that he hoped would enable a machine to perform most of the functions of a human brain and reproduce itself. Sadly, he died of cancer aged fifty-three before he could complete it.

Claude Shannon, another genius who was responsible for creating information theory and cryptography, and who coined the term 'bit' for binary digit, was also deeply involved in the earliest stages of AI. Shannon created a robot mouse that could learn to find its way through a maze, and a computer program that played chess, and in his later years he created other bizarre inventions such as a robot that could juggle balls. In 1955 Shannon and pioneers John McCarthy, Marvin Minsky and Nathaniel Rochester proposed a summer workshop to gather together scientists and mathematicians for several weeks to discuss AI. The Dartmouth Workshop was held for six weeks in the summer of 1956 and was the first ever focused event to explore (and name) AI. The weeks of discussion resulted in some of the key ideas that were to dominate this new field of research for many decades to follow.

A Proposal for the Dartmouth Summer Research Project on Artificial Intelligence, 31 August 1955

JOHN MCCARTHY, MARVIN L. MINSKY, NATHANIEL ROCHESTER AND CLAUDE E. SHANNON

We propose that a two-month, ten-man study of artificial intelligence be carried out during the summer of 1956 at Dartmouth College in Hanover, New Hampshire. The study is to proceed on the basis of the conjecture that every aspect of learning or any other feature of intelligence can in principle be so precisely described that a machine can be made to simulate it. An attempt will be made to find how to make machines use language, form abstractions and concepts, solve kinds of problems now reserved for humans, and improve themselves. We think that a significant advance can be made in one or more of these problems if a carefully selected group of scientists work on it together for a summer.

The rise and fall of AI

Excitement about AI grew rapidly in the years following the Dartmouth Workshop. New ideas about logic, problem solving, planning, and even simulating neurons fuelled researchers' optimism. Some researchers felt that machine translation would be solved very quickly, because of advances in areas such as information theory and new rules that described how words are put together in sentences within natural languages. Other researchers were investigating how the brain used neurons, connected together as networks, to learn and make predictions. Walter Pitts and Warren McCullough developed one of the first neural networks; Marvin Minsky designed the SNARC (Stochastic Neural Analog Reinforcement Calculator) – a neural network machine. By the early 1960s even highly experienced and intelligent pioneers were making slightly unrealistic predictions, given the current state of technology.

> **‘In principle it would be possible to build brains that could reproduce themselves on an assembly line and which would be conscious of their own existence.’**
>
> FRANK ROSENBLATT
> AI pioneer in perceptrons (1958)

Fuelled by such excitement, funding blossomed and researchers feverishly worked on machine translation and ‘connectionist’ (neural network) projects. But the hype was too much. By 1964, funders in the US (the National Research Council) were starting to worry about the lack of progress in machine translation. The Automatic Language Processing Advisory Committee (ALPAC) examined the issues. It seemed that researchers had underestimated the difficulty of word-sense disambiguation – the fact that the meaning of words depends on their context. The result was that the 1960s AIs made some rather embarrassing errors. Translated from English to Russian and back to English, 'out of sight, out of mind' became 'blind idiot'.

> **'Within our lifetime machines may surpass us in general intelligence.'**
>
> MARVIN MINSKY (1961)

The ALPAC report concluded that machine translation was worse than human translation, and considerably more expensive. After spending $20 million, the NRC cut all funding in response to the report, ending machine translation research in the US. At the same time, connectionist research was fading as researchers struggled to make the simple neural networks do anything very useful. The final nail in the coffin for neural networks was the book *Perceptrons* published in 1969 by Marvin Minsky and Seymour Papert, which described many of the limits of the simple neuron model. This spelled the end of neural network research. But it got even worse. Next came the Lighthill Report in 1972, commissioned by parliament to evaluate the progress of AI research in the UK. Mathematician Sir James Lighthill provided a devastating critique: 'Most workers who entered the field around ten years ago confess that they then felt a degree of naive optimism, which they now recognize as having been misplaced ... achievements from building robots of the more generalized types fail to reach their more grandiose aims.' The effects of the report had repercussions throughout Europe and the world. DARPA (Defense Advanced Research Projects Agency) cut its AI funding as the agency also realized that promised results were not being delivered in areas such as speech understanding. In the UK AI funding was discontinued in all but three universities (Essex, Sussex and Edinburgh). AI and intelligent robots had been completely discredited. The first AI winter had arrived.

Despite being deeply out of favour, a few AI researchers continued their work for the next decade. Earlier work was not lost; many advances simply became part of mainstream computer technology. Eventually, by the 1980s there was a new breakthrough in AI: expert systems. These new AI algorithms captured the knowledge of human experts in their rule-based systems and could perform functions such as identifying unknown molecules or diagnosing illnesses. AI languages designed to enable this kind of AI were developed, such as Prolog and LISP, and new specialized computers were built to run these languages efficiently. Soon, expert systems were being adopted in industries around the world and business was booming. Funding was now available again for AI researchers. In Japan, $850 million was allocated for the Fifth Generation computer project, which aimed to create supercomputers that ran expert system software and perform amazing tasks such as holding conversations and interpreting pictures. By 1985 more than $1 billion was being spent for in-house AI departments, and DARPA had spent $100 million on ninety-two projects at sixty institutions. AI was back, and with it came the overexcitement and hype once again.

> **"The time at which we might expect to build a computer with the potential to match human intelligence would be around the year 2017."**
>
> DAVID WALTZ
> AI pioneer on reasoning (1988)

But it didn't last. The power of conventional computers quickly overtook the specialized machines and the AI hardware companies went bust. Next it was discovered that the expert systems were horribly difficult to maintain and were prone to serious errors when given faulty inputs. Promised capabilities of AI were not achieved. Industry abandoned this new technology and funding quickly dried up once again. The second AI winter had begun.

The rebirth

Once again, despite being deeply out of favour, some AI research continued. In the 1990s, even the term AI was associated with failure, so it went under other guises: Intelligent Systems, Machine Learning, Modern Heuristics. Advances continued, and successes were simply absorbed into other technologies. Soon, a quiet revolution began, with more advanced fuzzy logics, new, more powerful forms of neural networks, more effective optimizers, and ever more effective methods for machine learning. Robotics technology also started to mature further, especially with new generations of lighter and higher-capacity batteries. Cloud-based computers made it possible to perform massive computation cheaply, and there was so much data being generated every day that the AIs had plenty of examples to learn from. Slowly, but with more and more vigour, AI and robotics returned to the world. Excitement grew yet again, and this time a little fear.

> **By 2029, computers will have human-level intelligence.**
>
> RAY KURZWEIL
> inventor and futurist (2017)

By 2019 it was the new AI summer, with thousands of AI start-ups worldwide busily trying to apply AI in new ways. All the major tech companies (Apple, Microsoft, Google, Amazon, Weibo, Huawei, Samsung, Sony,

> **We should not be confident in our ability to keep a super-intelligent genie locked up in its bottle for ever.**
>
> NICK BOSTROM
> head of Future Of Humanity Institute, Oxford (2015)

> **Most executives know that artificial intelligence has the power to change almost everything about the way they do business – and could contribute up to $15.7 trillion to the global economy by 2030.**
>
> PRICEWATERHOUSECOOPERS (2019)

IBM – the list seems endless) were together investing tens of billions of dollars in AI and robotics research. For the first time, AI-based products were being sold to the public: home hubs that recognized voices, phones that recognized fingerprints and faces, cameras that recognized smiles, cars that automated some driving tasks, robot vacuum cleaners that cleaned your home. Behind the scenes, AI was helping us in a hundred tiny ways: medical scanners that diagnosed illnesses, optimizers that scheduled delivery drivers, automated quality-control systems in factories, fraud detection systems that noticed if your pattern of spending changed and stopped your card, and fuzzy-logic rice cookers to make perfect rice. Even if we once again decided not to call it AI in the future, this technology was too pervasive to disappear.

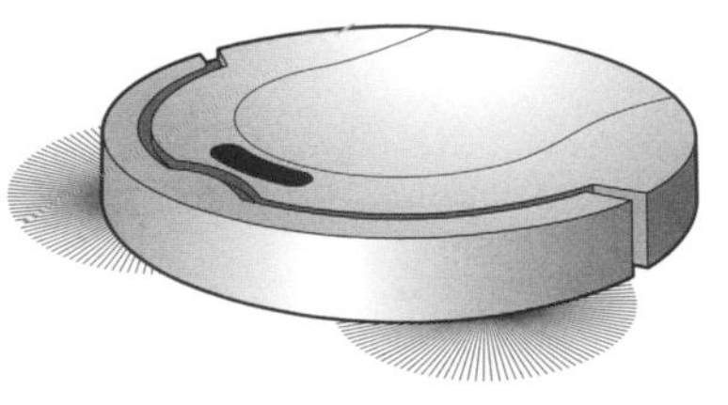

There had never been so much excitement, so many researchers, so much money, so much hysteria. Despite the ups and downs

> **Success in creating AI would be the biggest event in human history. Unfortunately, it might also be the last, unless we learn how to avoid the risks.**
>
> STEPHEN HAWKING (2014)

of AI's popularity, progress in research had never stopped. Today is the culmination of thousands of years of effort poured into some of the most miraculous technologies humans have ever created. If there has ever been a golden age for AI, it is now. Our extraordinary intelligent technology doesn't just help us, it reveals to us the very meaning of intelligence, while posing deep philosophical questions about what we should allow technology to do. Our future is intimately tied to these smart devices, and we must navigate the minefields of hype and misplaced trust, while learning how to accept AI and robots into our lives.

Each of the following chapters of this book will show you some of the most extraordinary AI inventions so far, and what they might mean for our future. Welcome to the world of AI and robotics.

02 CHOOSE THE RIGHT PATH

'I never guess. It is a shocking habit destructive to the logical faculty.'

ARTHUR CONAN DOYLE

However of the clock for intuitionists
When of proof, though I not yet, being mine,
Or true a certain portion mathematical
And its element asserted: this was.

It was born of the
Axiom-schemas that now I
Not yet fear from implication.

One of the concepts of youth, the eye. When
That, there's a mathematics eternal.

Perhaps this is not the greatest poetry in the world, but this short collection of quatrain, haiku and couplet was generated in a split second by an AI, which was attempting to express ideas about logic with a flavour of the sonnets of Shakespeare. When we read such poems, we might find some deeper meaning in the words. Somehow the AI captures something that makes us wonder if there is a message being communicated.

In this case, unfortunately, there is not. The poems were generated by a computer following a set of rules that define the structure of each type of poetry. (For example, a Haiku comprises three unrhymed lines of five, seven, and five morae, while a couplet comprises two lines which may be rhymed or unrhymed.) The words were randomly picked from source text (several paragraphs, which included a history of logic, a sonnet from Shakespeare, and a section of text from a 1927 von Neumann paper on logic). Use different source texts and different rules, and the AI will churn out poems about anything you like, in whatever style you have defined.

Symbolic AI

In symbolic processing, words are treated as symbols that relate to each other according to a set of rules. It is almost as though words are objects that you can move around and manipulate, perhaps transforming them, in the same way that the rules of mathematics allow us to manipulate numbers. Symbolic AI allows computers to think using words.

It's perhaps not surprising that symbolic AI was one of the first and most successful forms of AI, because it was built upon the new ideas of logic that had been developed a few decades earlier. Towards the start of the twentieth century, mathematicians such as Bertrand Russell, Kurt Gödel and David Hilbert had been exploring the limits of mathematics to see if it was possible to prove everything, or whether some things that you could express in maths were actually unprovable. They showed us that all of mathematics could be reduced to logic.

> **Thought was still wholly intangible and ineffable until modern formal logic interpreted it as the manipulation of formal tokens.**
>
> ALLEN NEWELL (1976)

Logic is a very powerful kind of representation. Anything expressed in logic must be true or false, allowing us to represent knowledge; for example: raining is true; windy is false. Logical operations allow us to express more complex ideas: if raining is true and windy is false, then 'use umbrella' is true. This little logical expression can also be given as a truth table:

raining	windy	use umbrella
false	false	false
true	false	true
false	true	false
true	true	false

RUSSELL'S PARADOX IN PREDICATE LOGIC

Consider this paradox by philosopher Bertrand Russell: 'There is a man who shaves all and only men who do not shave themselves'. It's a paradox because if the man shaves himself, then he can't shave himself according to the rule. But if he doesn't shave himself, then he must shave himself according to the rule. We can turn this into a complicated-looking symbolic logical expression like this:

$$(\exists x)\Big(\text{man}(x) \wedge (\forall y)\Big(\text{man}(y) \rightarrow \big(\text{shaves}(x,y) \leftrightarrow \neg\text{shaves}(y,y)\big)\Big)\Big)$$

Don't be scared! A literal translation into English gives us: 'There exists something called x that is a man and for all things called *y* where *y* is a man then x shaves *y* if and only if not *y* shaves *y*.' It's useful, because with this kind of predicate logic, it is possible to make proofs. In this case, you can reveal the paradox by asking, 'Does the man shave himself?' Or, in the logic expression, what do you get if $x = y$? Substitute x for *y* and the result is that shaves (x, x) and its inverse $\neg$shaves (x, x) are true. In other words, the man must shave himself and he cannot shave himself at the same time – a paradox. (Russell showed that mathematics is incomplete using a paradox similar to this one, i.e., that it is not possible to prove everything in mathematics.)

When we prove something in maths we show that the assumptions logically guarantee the conclusion. Maths is built on such proofs. So if we have the assumptions, 'All men are mortal', and 'Socrates is a man', we can prove that 'Socrates is mortal.'

Predicate logic, a more complicated and commonly used type of logic, even allows normal sentences to be turned into a kind of logic notation (known as formal logical expressions).

Logic was considered so powerful that the early pioneers of symbolic AI decided that symbolic logic was all that you needed for intelligence.

This belief was based on the idea that human intelligence was all about manipulating symbols. These researchers argued that our ideas about the world around us are encoded in our brains as symbols. The idea of a chair and a cushion could be encapsulated by the symbols 'chair' and 'cushion' and abstract rules such as 'a cushion may be placed on a chair', and 'a chair is not placed on a cushion'.

> **A physical symbol system has the necessary and sufficient means for general intelligent action.**
>
> ALLEN NEWELL AND HERBERT A. SIMON (1976)

Chinese room

But some philosophers didn't agree. They argued that manipulating symbols is quite different from understanding what the symbols mean. John Searle was one such philosopher, who neatly described his objection in the form of a story about a Chinese room. He imagined himself inside the room, and every so often a slip of paper with Chinese characters was handed to him via a slot. He would take the paper, and match the symbols against others inside rows of filing cabinets in order to look up an answer, which he would carefully copy down on a new piece of paper and return.

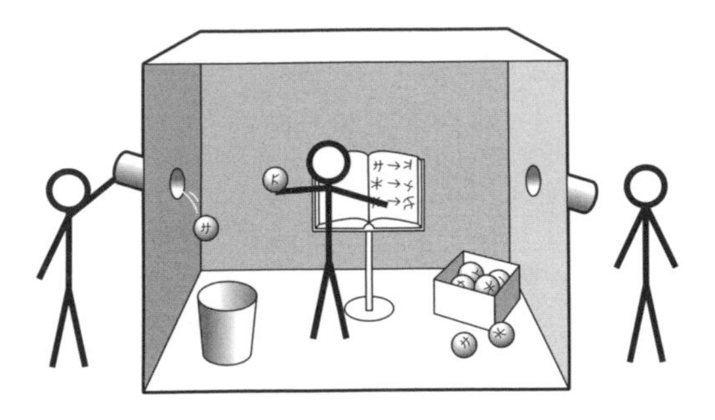

From the outside, it appeared as though you could ask any question and receive a sensible answer. But on the inside, there was no understanding. At all times Searle was following rules, using symbols to look up other symbols. At no time did he ever understand what was going on – for he could not read or understand Chinese.

Searle argued that this is exactly what an AI is doing when it performs symbolic processing. It manipulates symbols according to rules, but it never understands what those symbols and rules mean. Given a question, 'What colour is a ripe banana?', the AI might be able to look up the answer and reply, 'yellow'. It might even follow other rules to make it sound more human and reply, 'Yellow, of course. Do you think I'm stupid?' But the AI doesn't know what 'yellow' means. It has no link from the symbol 'yellow' to the outside world – for it doesn't know about and never experiences the outside world. Such an AI cannot have intentionality – the ability to make a decision based on its understanding. So, Searle argued, AI is merely simulating intelligence: 'The formal symbol manipulations by themselves don't have any intentionality; they are quite meaningless,' he explains. 'Such intentionality as computers appear to have is solely in the minds of those who program them and those who use them, those who send in the input and those who interpret the output.'

> **No single logic is strong enough to support the total construction of human knowledge.**
>
> JEAN PIAGET
> psychologist

Even if it passed the Turing test, it wouldn't matter. AI is a machine designed to fool us, just like the automata of ancient Rhodes. AI is weak, and a 'strong AI' that has real intelligence might be impossible.

Searching logic

Despite the criticism, the ideas of symbolic processing had already shown considerable success. Back in 1955, Newell, Simon and Shaw developed the first AI program ever (even before the term artificial intelligence had been introduced). They called it the Logic Theorist, and at the Dartmouth Workshop in 1956, they proudly showed it off to the other researchers. Using logical operations, the program was able to prove mathematical formulae. To demonstrate, Newell and Simon went through the popular maths book *The Principia Mathematica* by Alfred Whitehead and Bertrand Russell and showed that it was able to prove many of the formulae, and in some cases produce smaller and more elegant proofs.

Several other versions of the Logic Theorist were created, one known as the General Problem Solver, in 1959. This AI was able to solve a range of problems, including logical and physical manipulation. The GPS had an important trick to enable it to work so well. It separated its knowledge (symbols) from the method used to manipulate those symbols. The symbols were manipulated using a piece of software called a solver, which used *search* to find the right solution.

Imagine you're a robot and you have to move a stack of differently sized discs from one pile to another, keeping them in order of ascending size. It's a game known as the Towers of Hanoi. You can only move one at a time

NEWELL, SIMON AND SHAW

Allen Newell was a computer scientist and cognitive psychologist based at the RAND Corporation and at Carnegie Mellon University. He worked with Simon and Shaw on the Logic Theorist project and together they created many fundamental inventions in AI. Newell also created the concept of processing lists, which later became an important AI language known as LISP. Programmer John Shaw created the idea of the linked list, a way of linking data that has been used in programming languages around the world ever since. In addition to the GPS (General Problem Solver), Herbert Simon helped develop AI chess-playing programs and made contributions in economics and psychology. He even wrote one of the first ever papers on emotional cognition, which he implemented as drives and needs that might occur in parallel, interrupting and modifying the behaviour of a program. Newell and Simon created an AI lab at Carnegie Mellon University and made many advances in symbolic AI through the late 1950s and 60s.

and you can only pick up the top disc from a pile. You cannot place a larger disc on top of a smaller one. And you can only have three piles. How do you move them? For each move, given the current state of the piles, you may have two or more moves available to you: do you choose this disc or that disc? Do you put the disc there or there? Once you've moved the disc, you now have more moves available to you, and when you have made that decision you have more options again. The game is like a tree of possibilities, each branch taking you towards the solution, if you pick correctly. But with so many choices, how to choose the right moves?

The solution is search. The AI imagines making choice after choice, following the tree of possible moves down as far as it has time to look, and makes a judgement – at this point, if I made these choices, do I appear to be closer to reaching the solution? Consider enough combinations of choices, and you will find a good path, which means you can now make the next decision – I will put this disc there. Once done, the AI can search the new tree of possibilities, one choice deeper, and figure out the next move.

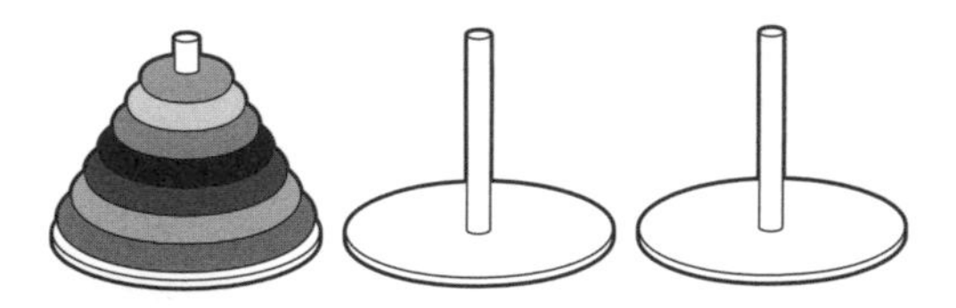

> **Anything that gives us new knowledge gives us an opportunity to be more rational.**
>
> HERBERT SIMON (2000)

Search combined with a symbolic representation of knowledge became a standard approach in AI. Whether the AI is figuring out how to win a game of chess or Go, or trying to prove a formula, or planning which route a robot might take to avoid a set of obstacles in its path, it is probably searching among thousands or millions – or trillions – of possibilities to achieve its goal. The sheer size of the search space was quickly a major limiting factor in search-based symbolic AI, so many clever algorithms were created to prune away unlikely-looking areas of the tree or divide the problem into smaller subproblems. Reduce the space and there's less to search.

But the size of the space (its combinatorics) still proved troublesome. Perhaps ironically, ideas such as the General Problem Solver were found to be impractical for general problems – the space of possibilities becomes intractable (not searchable in a practical amount of time). While these AIs could solve 'blocks world' problems such as the Towers of Hanoi, they struggled with the complexities of the real world. Instead, better success was found if each AI focused on a specific topic. Create a set of rules about which medical symptoms are associated with each illness, and the computer can then ask a series of questions – 'Do you feel pain?', 'Is it a sharp pain or dull ache?', 'Where is the pain located?' – in order to match symptoms and suggest one or more potential illnesses. These AIs became known as expert systems and for a time were extremely popular (perhaps a little too popular, as we saw in the previous chapter). Although larger expert systems suffered from maintenance problems, such focused expert systems remain in use for medical diagnosis, support systems for automobile engineers, fraud detection systems and interactive scripts for salespeople.

Storing knowledge

Many ideas in symbolic AI are to do with how best to represent information, and how to use that information. Rules and structured 'frames' have merged with object-oriented programming languages, and there are many powerful ways that knowledge can be stored; for example, using inheritance, such that a parent object 'tree' might contain the children 'oak' and 'birch', or message passing, such that an object 'seller' might send an argument '10 per cent discount' to trigger a behaviour in another object 'price'. Entire knowledge representation languages have been created, sometimes called ontologies, with their own complicated structures and rules. Many are based on logic, and may be combined with automated reasoning systems to enable them to deduce new facts, which can then be added to their knowledge, or to check the consistency of existing facts. For example, say the AI has learned that 'bicycle' is a type of 'pedal-powered transportation' that makes use of 'two wheels'. If 'tandem' is a type of 'pedal-powered transportation' that also has 'two wheels' then the AI could easily deduce the new fact: 'tandem is a kind of bicycle'. But a different type of 'pedal-powered transportation' with 'zero wheels' doesn't fit the rule, so the AI will conclude that 'pedalo boat' is not a 'bicycle'.

With the growth of the internet, compiling vast collections of facts became easier and easier. Several major efforts in general artificial intelligence aimed at combining enough knowledge together that an AI could start to help us in multiple areas. Cyc was one such effort, which has now been compiling common-sense facts and relationships into a giant knowledge base for several decades.

One computer scientist took this idea even further. William Tunstall-Pedoe created True Knowledge, a vast network of knowledge provided by users on the internet that comprised more than 300 million facts. In 2010 Tunstall-Pedoe decided that since his AI knew so much, he would ask it a question that no human could hope to answer. 'It occurred to us that with over 300 million facts, a big percentage of which tie events, people and places to points in time, we could uniquely calculate an objective answer to the question "What was the most boring day in history?"'

True Knowledge looked at all the days it knew about from the start of the twentieth century and decided that the answer was 11 April 1954. On this day, according to the AI, Belgium held a general election, a successful Turkish academic was born, and the footballer Jack Shufflebotham died. These were fewer events compared to all the other days, and so the AI decided that this was the most boring day. True Knowledge eventually became Evi, an AI that you could speak to and ask questions. In 2012 Evi was acquired by Amazon and became Amazon Echo, the well-known household talking AI.

Symbolic AI is today growing as the internet grows. While AIs such as Cyc and Evi relied on thousands of users providing concepts manually, Sir Tim Berners-Lee, the creator of the World Wide Web, has long pushed the idea that the WWW should become a GGG (giant global graph) of concepts. Instead of just building websites for people to use, the websites should also hold data in a form that computers can understand. Websites were traditionally built like documents, with text, images and videos, or like programs, with behaviours that are triggered when forms are filled and buttons pressed. In Tim's dream, inside every web page concepts should be labelled with names and unique identifiers. In the Semantic Web, as it is known, websites become databases of concepts, where every element is an object in its own right that can be found independently and has a clear textual label or type. If the entire WWW became a GGG, then our AIs would be able to search, make deductions and reason about the world's knowledge.

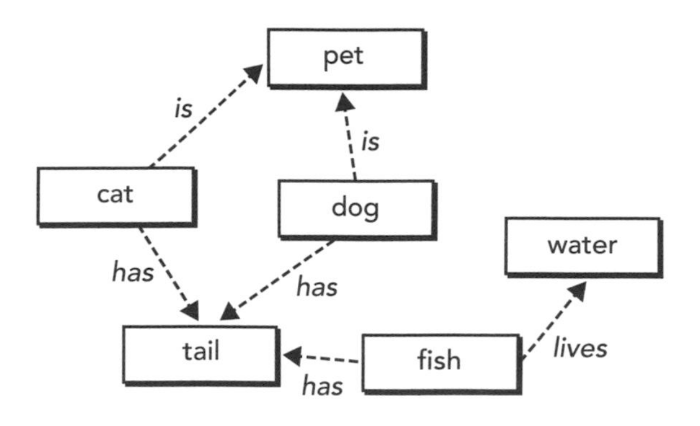

This grand dream for symbolic AI has sadly not been adopted by most web developers in the world, who continue to put vast quantities of data online in a form that AIs struggle to recognize. And the need is becoming urgent. In 2019, it was estimated that 80 per cent of new data will be unstructured – that's without a consistent representation for knowledge that computers can understand, such as textual documents, images and videos. (Think about all those emails or reports you write as 'free text' – that means without explicitly chunking them into labelled sections. Or all the photos and videos you take with your phone – you're not going through and labelling every scene, or item in frame.) At the same time, the amount of data is growing year by year. By 2019 there were 4.4 billion internet users, an increase of 80 per cent in five years, and 293 billion emails sent daily. There were 40,000 searches per second of the internet with Google and 7,800 tweets per second on Twitter. More and more companies used the internet as part of their business, and generated their own huge quantities of data. In 2016 we were generating 44 billion Gb per day worldwide. It has been estimated that by 2025 we will generate 463 billion Gb per day.

> **I have a dream … Machines become capable of analysing all the data on the web … the day-to-day mechanisms of trade, bureaucracy, and our daily lives will be handled by machines talking to machines, leaving humans to provide the inspiration and intuition.**
>
> TIM BERNERS-LEE (2000)

We no longer have a choice – no human can make sense of these mind-boggling quantities of data. Our only hope is to use AI to help us. Luckily, as we will see in later chapters, other forms of AI are now able to process unstructured and unlabelled data, and tag them with symbolic labels, giving the symbolic AIs what they need in order to think about them. In the end, maybe it doesn't matter if this is true intelligence (strong AI) or simply a kind of 'pretend intelligence' (weak AI). Processing networks of symbols according to rules enables our computers to make sense of our vast universe of data. And perhaps one day, Berners-Lee's dream for the web will come true.

03 WE ALL FALL DOWN

'You don't learn to walk by following rules. You learn by doing, and by falling over.'

RICHARD BRANSON

We see a grainy film footage. A strange wheeled robot looking rather like a wobbly photocopier on wheels with a camera for a head is trundling around a space occupied by large coloured cubes and other simple shapes. In the background we hear the mellow jazzy sound of 'Take Five' played by The Dave Brubeck Quartet. A narration begins, with a high-pitched whine in the background:

> At SRI we are experimenting with a mobile robot. We call him Shakey. Our goal is to give Shakey some of the abilities that are associated with intelligence – abilities like planning and learning. Even though the tasks we give Shakey seem quite simple, the programs needed to plan and coordinate his activities

> are complex. The main purpose of this research is to learn how to design these programs so that robots can be employed in a variety of tasks ranging from space exploration to factory automation.

This was the state of the art in robotics in 1972. Shakey (who had a brain based elsewhere in a massive mainframe computer) was able to use its camera to identify the simple objects around it, build a model of its simple world, and plan where to go and what to do, making predictions about how its actions would modify its internal model. Shakey wasn't very fast or very clever, but he represented a revolution in AI research. For the first time, AI could enable a robot to navigate and perform actions in the world (albeit in a very neat environment).

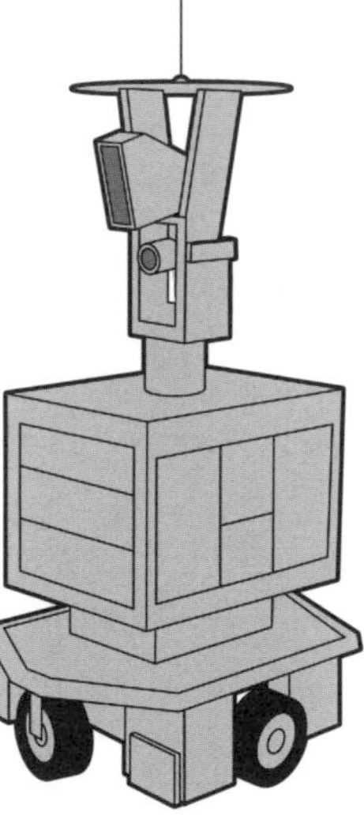

It was a great start, but it didn't work very well. All the planning and decision making took a large amount of computation power, and, combined with the limited vision systems of the day, Shakey and robots like him were slow, unreliable and unable to cope with noisy, real-world environments. This was the accepted way to make intelligent robots, but researchers were

ROBOTS IN THE SIXTIES

Robots are troublesome to build. It's hard enough making them move around, but the problems related to control and sensing are where we need artificial intelligence to help. Back in 1960, when AI was not really up to the job, robots were quite terrifying. Hardiman was one example, developed between 1965 and 1971 by General Electric. Intended as a powered exoskeleton for a human (inspiration for the exoskeleton worn by Ripley in the movie *Aliens*), the suit only achieved violent uncontrolled motion, and the developers were never able to make more than a single arm work. The Hopkins Beast was more successful – it was a complex version of Grey Walter's Elsie robot, controlled by a number of early transistors, enabling it to wander randomly through the halls of its home in Johns Hopkins University and plug itself into wall outlets for power. The Walking Truck was another robot pioneered at General Electric in 1965. This massive machine was intended to carry equipment over rough terrain. There was no computer control of its movement – it required a skilled human operator to control its four metal legs, using their own arms and legs.

finding it harder and harder to make progress. In robotics, the neat and logical way of thinking started to be challenged by the need for something messier. Such challenges led to the formation of two main camps in AI research: the 'neats' and the 'messies'.

Elephants don't play chess

While the 'neat' researchers preferred carefully designed and often mathematically provable methods, the 'messies' claimed that such methods did not scale beyond the artificial blocks world (like the Towers of Hanoi game). If you're trying to create a robot that can move around and understand its world, then logic and an assumption that the environment is perfect results in failure. Rodney Brooks, the founder of the iRobot company and creator of the iRoomba robot vacuum cleaners, summarized his critique in a seminal article entitled 'Elephants Don't Play Chess'. He argued that AI's focus on logical game-playing had nothing to do with intelligent behaviour in the real world. Being able to play a good game of chess does not help you to walk, avoid obstacles or cope with the ever-changing nature of the real world. A robot should not build logical internal models comprised of symbols, make a plan by manipulating and searching those symbols, and then use the result to determine its behaviour. Instead, to achieve practical robotics, we should create AIs that are physically grounded.

Brooks was a practical scientist, and after many years' experience building robots, he had found a different approach. He argued that the world is its own best model, so we should let the world directly affect the behaviour of the robot without any symbols at all – we should connect perception to action.

> **'We argue that the symbol system hypothesis upon which classical AI is based is fundamentally flawed.'**
>
> RODNEY BROOKS (1990)

It was an idea first explored by Grey Walter and his robot tortoises, as we saw in the first chapter. Brooks called his version the 'subsumption architecture'. Using this idea, a series of simple modules control robot behaviour, each one interrupting the other if its needs become more urgent. One module might be in charge of moving the robot towards a target, another might be in charge of avoiding obstacles. The first would take priority until something unexpected got in the way, triggering the second module to take avoiding action. Brooks represented the behaviours using finite state machines.

Finite State Machines (FSMs) are a common kind of 'brain' for robots. They work by identifying a series of 'states' that the robot can be in. For example, a very simple robot might have three states: moving randomly, moving forwards, and turning. It can transition from one state to another when it senses certain things. So every time it senses a target, it would switch to (or maintain) the state of moving forwards. Whenever it senses an obstacle in front, it switches to (or maintains) the state of turning. If it senses nothing, it switches to (or maintains) the state of moving randomly – see diagram. This generates a simple kind of architecture where the robot randomly wanders about, avoiding obstacles, until it finds a target destination. Add additional FSMs connected to the same sensors and effectors, and make some take precedence over others depending on the sensors and states, and you have a subsumption architecture.

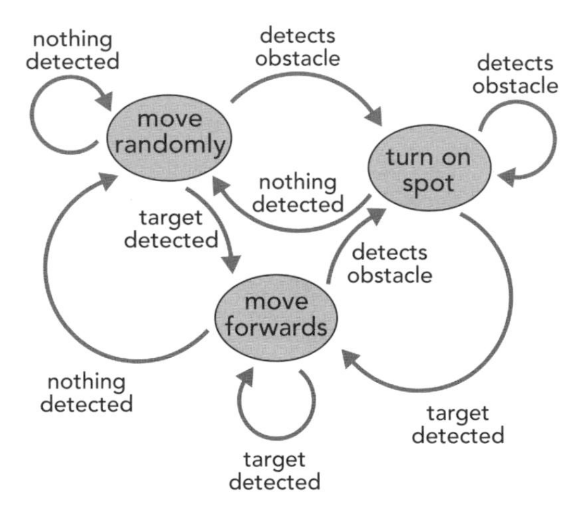

As Brooks later explained: 'If I'm trying to get somewhere quickly, I'm not consciously thinking about where to place my feet. I've got a different layer that does the foot movements. I've got separate processes running. That was the idea with the behaviour-based approach.' Brooks' six-legged robot Genghis used fifty-seven FSMs in combination.

The approach resulted in a lightweight, quick-to-run AI that enabled robots to do more, using less computation, than had ever been achieved previously. Brooks demonstrated the effectiveness of his approach through numerous projects (and companies) that pioneered many different types of robot, including the Sojourner lunar robot that explored the surface of the moon for several weeks in 1997 using this form of behavioural control.

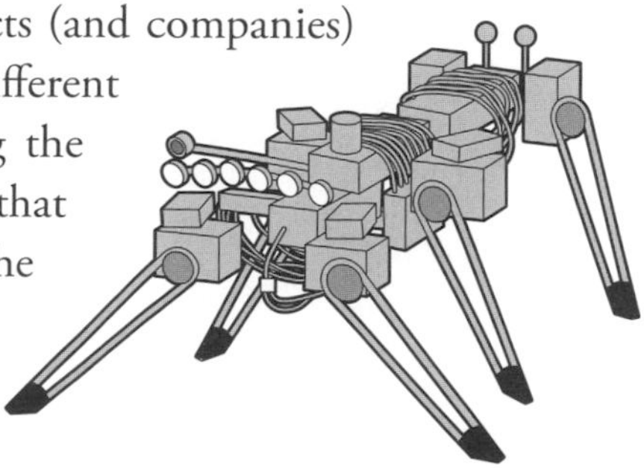

Well-behaved robots

Eventually the subsumption architecture was simplified from a mixture of finite state machines into behaviour trees – a rather more elegant way of representing the same concepts. These were adopted by the computer games industry in order to drive the behaviour of 'virtual robots' – all the aliens, monsters and other characters in computer games that challenge us. Unity and Unreal are two of the biggest software platforms used to create two-thirds of the computer games by 2019. Both platforms use behaviour trees.

SOJOURNER PATHFINDER ROVER

Sojourner was a Mars Pathfinder rover robot, the first to rove around exploring the surface of another planet. An 11.5 kg robot with six wheels, three cameras and solar cells on its upper surface to generate power, it first touched down on 4 July 1997. Its usual method of control was by a human operator, wearing a 3D headset to watch the robot from its base station. Technology was a little simpler in the 1990s, so the robot's computer processor only ran at 2 MHz (a thousand times slower than today's computers) and had only 64 K of memory (less than a ten-thousandth of today's computers). It didn't even have rechargeable batteries, so once its batteries went flat it could only operate during the day, powered by its solar panels. The robot was so far from Earth that there was a time delay or a 'round-trip' time of twenty minutes (the operator sends a signal and the robot sends back a reply). This meant that it was vital for it to have its own autonomous control, in case it drove itself off a cliff or into a rock during that time. The subsumption architecture enabled it to perform navigation, hazard detection and avoidance all by itself.

Lightweight, ultra-fast control modules that connect perception to action and kick in when they're needed are now the cornerstone of practical robotics. The robot company Boston Dynamics provides some remarkable examples of just how effective this type of control can be, especially when combined with more springy actuators (the part of the robot that produces movement) that resemble the movement of muscles. Their robot dogs and bipedal robots can withstand being kicked and still manage to retain their balance, thanks to the clever control systems (which are also combined with other AI approaches such as planners and optimizers).

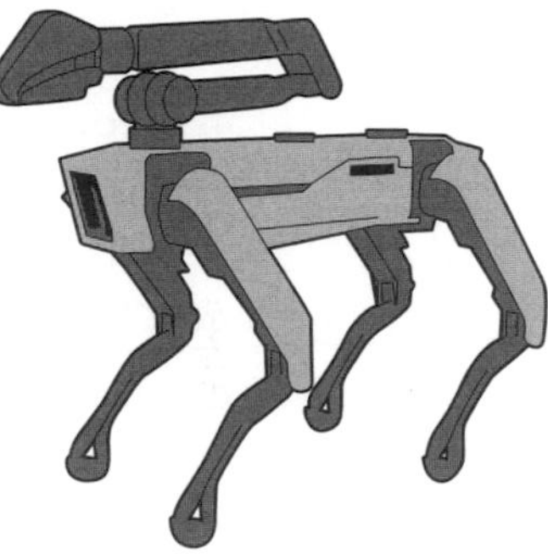

With such amazing technology, ever better actuators, sensors, batteries and AIs, surely we must be about to have human-like robots who can help us and work with us in our homes and workplaces? Not really. AIs and robots are clever but they fail one simple test. No robot with an AI brain can walk around reliably in your home without hitting something or falling over. The action of simply walking around without tripping may sound much less difficult than understanding speech. In fact, it's much harder. Controlling robots in unpredictable environments remains one of our biggest challenges, and it's caused by a number of factors. To be able to move fluidly and with grace, a robot needs more and more effectors (motors, pneumatic pistons, or other muscle-like actuators) and it needs more and more sensors. But more effectors in noisy environments mean that you create a chaotic, unpredictable control problem, and more sensors mean an overwhelming amount of data that needs to be processed and understood – with serious time limits, because if you take too long to figure out where a limb is, or you incorrectly place it, then before you know it you're on the floor. So, most robots with legs still fall down. A lot.

Today, and for the foreseeable future, the most effective robots are not humanoid – they are the robots whose form perfectly suits their function. Our factories are filled with robot arms that are programmed to assemble many of our mass-produced products. The latest operating theatres contain several medical robotic machines to assist with complex surgeries and life support. Your washing machine is a robot. Your central heating system or air-conditioning unit is a

> **What did everyone think robot vacuuming was going to be? Well, they think Rosie the Robot from *The Jetsons*, a human robot that pushed a vacuum. That was never going to happen.**
>
> COLIN ANGLE
> CEO of iRobot

robot. And while they cannot walk around your home, we do have robot vacuum cleaners, which mostly work without getting stuck too often.

Self-driving vehicles

Perhaps the most exciting form of AI-driven robot that has started to become a reality in the last few years is one that actually transports us. Autonomous cars were first demonstrated back in the 1980s, with several American projects achieving cars that drove themselves for several thousand miles and could drive during both the day and night. While successful, computer vision was still primitive and so – despite significant funding from DARPA, the US Army and Navy – real breakthroughs took place only when methods such as deep learning transformed the capabilities of AI to process camera and LIDAR (3D laser scanning) systems. Many organizations (Tesla, Waymo, Uber, General Motors, Ford, Volkswagen, Toyota, Honda, Tesla, Volvo and BMW) are investing heavily in the technology – indeed by 2019 more than forty companies were developing their own autonomous vehicles. With the AIs now able to make more sense of their messy surroundings, self-driving vehicles can handle many driving situations from the simple automatic braking to avoid collisions, to the more complex automated parking, to even achieving full autonomy as a taxi service in controlled environments such as New York's 300-acre Brooklyn Navy Yard. The potential for the technology is amazing, but there are a considerable number of issues arising as such products become available.

Self-driving cars are not clever enough to be fully autonomous at all times. While AIs may now be able to recognize shapes such as other vehicles or pedestrians, they lack judgement and understanding of the context of what they perceive and so cannot hope to match the abilities of a good driver. No self-driving car is currently capable of full automation; all require constant supervision by a human driver, who may need to take control if the AI becomes confused. This may not be understood fully by those who own such vehicles, a fact that has resulted in some fatalities. Even if drivers do understand, the ability to remain alert and ready to take back control in a split second is not easy to master or maintain. A new kind of driving test might be required. Liability issues also arise in the case of accidents. If you were not driving your car when it damaged another vehicle, are you to blame, or should the manufacturer of the AI driver of your car become

> **Users were less accepting of high autonomy levels and displayed significantly lower intention to use highly autonomous vehicles.**
>
> AUTONOMOUS VEHICLE ACCEPTANCE MODEL SURVEY (2019)

liable? If a fully autonomous taxi caused a road injury or death, one certainly could not sue the passengers of the vehicle.

Robots in our society

Autonomous vehicles highlight one of the more difficult aspects of the introduction of robots to our societies. Society is inevitably affected and will be changed. By releasing a technology that automates the skill of driving, we de-skill ourselves, making human drivers less capable instead of more, and potentially making roads less safe. And if fully autonomous cars are perfected in a few more years, will we then see teenagers playing 'stop the car' by sticking out a leg and fooling the AI into assuming they're about to cross the road? Will they become the terrorists' tool of choice to cause havoc, by hacking them? There's also the impact to jobs, as drivers might be replaced – a fear that is shared by factory workers.

> **Up to 20 million manufacturing jobs around the world could be replaced by robots by 2030.**
>
> OXFORD ECONOMICS (2019)

Consultants predict that these new robot technologies will hit lower-income regions the most, with many less-skilled jobs being lost. However, the news is not all bad. Analyses show that the faster countries adopt these robot technologies, the faster the short- and medium-term growth of their overall economies, resulting in the creation of more jobs.

> **Will robots inherit the earth? Yes, but they will be our children.**
>
> MARVIN MINSKY (1994)

Ultimately, although AI and robots seem frightening, this is nothing more than just another new technology, and humans have been making new technologies for thousands of years. Every time we invent something new, we may sadly make some jobs that depend on the previous technology obsolete. But every new creation can result in an entirely new industry. While factory workers may no longer be needed to assemble so many products, more people will be needed to build, maintain and program robots. While the number of taxi or truck drivers could decrease, more jobs will be created in the construction industries to ensure road infrastructure is suitable for autonomous vehicles, and in the manufacture and servicing of these considerably more complex vehicles. (Not to mention all the lawyers needed to resolve the tricky new lawsuits.)

> **The key question we should be asking is not when will self-driving cars be ready for the roads, but rather which roads will be ready for self-driving cars.**
>
> NICK OLIVER
> Professor at the University of Edinburgh Business School (2018)

Developing AIs that can control robots, whatever those robots may look like, remains a hot topic for research. There are many unresolved technical problems, and it will be many decades before unsupervised AIs in uncontrolled environments are safe enough for us to entrust them with our lives. It should perhaps also always remain a choice whether we wish this to happen. The robots are coming, but how we choose to accept our creations is up to us.

04 FIND THE RIGHT ANSWER

'There's a way to do it better – find it.'

THOMAS EDISON

We see a strange snake-like form, somehow made of cube-like blocks, yet swimming in an undulating motion through water. Now we see three blocky tadpole-like forms, swimming together with a smooth grace that belies their LEGO-like construction. A turtle-like creature swims into view, made of just five rectangular blocks – one for the body and one for each flipper. Somehow it swims purposefully towards its goal, manoeuvring expertly around in the water like a hunter tracking its prey.

These are the evolved virtual creatures of computer graphics artist and researcher Karl Sims – the work that has inspired hundreds of scientists since he first released it back in 1994. His menagerie of swimming, walking, jumping and competing creatures astonished the scientific community. While their virtual bodies might have been relatively simple collections of blocks, their artificial brains were highly complex networks of maths functions and operations that used sensor inputs and produced intelligent-looking motion and behaviours. They moved in perfectly simulated virtual worlds, with simulated water that they could use to swim within, or simulated ground, gravity and the laws of physics, enabling them to walk, run or jump.

But this was not enough to astonish the other scientists. What was truly groundbreaking was the fact that Sims didn't program these creatures. He didn't design any of them. He didn't make their bodies, and he didn't create their brains. He was as amazed as everyone else when he first saw them. Sims had evolved his virtual creatures.

Evolving artificial life

Sims used a genetic algorithm to evolve his virtual creatures, where his quality measure (or 'fitness function') was how far they could swim, or walk, or jump – the further the better. To solve this problem, his genetic algorithm evolved both the bodies and brains of virtual creatures.

Sims didn't even know how these solutions worked. But he could see that they did. When describing his work to an amazed audience of the International Conference on the Simulation of Adaptive Behaviour in 1994, he explained how complex the brains of the creatures had become. The turtle-like creature may have had a body comprising just five simple blocks, but if its brain were printed out, the length of the paper would stretch across a significant part of the large conference auditorium. 'It allows us to go beyond what we can design. If I were to try to hook together these sensors, neurons and effectors myself, then I might never come up with a good solution, but evolution can still do it.'

Evolution in a computer may sound bizarre, but it's an AI approach that has been around since the early days of computers. Rather than trying to write a program that solves a problem by performing a calculation and outputting the answer, in evolutionary computation practitioners create a virtual world and let the computer find the solution all by itself by breeding better and better solutions. The genetic algorithm is one such approach. It works by creating a random population of rather useless solutions, ranking them in order of fitness (how well they solve the problem) and letting only the fitter solutions have children. The new generation of solutions is then ranked in order of fitness, the fittest of these have children again, and so on. Every time solutions have children,

WILLIAM LATHAM (1961–)

In 1983, Latham was a young British artist with unusual ideas. He was fascinated with the natural world and the complex forms of living creatures. He started developing his own style, drawing out vast family trees of imaginary forms, lineages of shapes that slowly changed over time according to his rules of inheritance and mutation. Following a talk he gave to the research lab at IBM Hursley, he was invited to become a research fellow, where he formed a long-lasting collaboration with IBM mathematician and developer Stephen Todd, and together they created the Mutator computer program. Latham commercialized the software in a variety of software releases, and even generated artwork for music album covers. Before long, his computer animations were regularly used at raves and dance clubs. Latham for a time had his own computer games company, which created several successful hits. Most recently he returned to his evolutionary art as a Professor at Goldsmiths College London, now working with Stephen Todd and his son, software developer Peter Todd. Together they created their own company, London Geometry, to take the ideas further.

the kids inherit digital genetic code from their parents, mixed up together so that each child has random chunks from each parent, with an occasional random mutation to introduce novelty. Let the GA run for enough generations and the population evolves highly fit solutions that solve the problem.

Sims wasn't the only pioneer to showcase the originality and novelty of digital evolution. Some five years earlier, artist William Latham and Stephen Todd developed Mutator. It was a revolutionary form of art for an artist to create – because strictly speaking, he didn't create it. Latham's art was all evolved in a computer. In this case, Latham acted like the 'eye of God' – it was his choice which solutions had children and which died, as he judged them for artistic merit. Like breeders of animals, Latham bred his art by selecting those he deemed worthy, and from the random chaos emerged extraordinary forms, swirling shapes and otherworldly images.

Some of the forms looked like they could be from an alien planet ... they're continually evolving, always subtly changing shape.

WILLIAM LATHAM (2015)

BIO-INSPIRED OPTIMIZATION

Genetic algorithms and their closely related cousins, evolutionary strategies and evolutionary programming, date back to the earliest days of computer science. Ant-colony optimization and artificial immune systems were rather more recent additions in the 1990s. But in the last few years, researchers have shown just how many natural processes can inspire optimization. There are optimization algorithms based on bees, natural processes such as central force optimization, the intelligent water drops algorithm and river formation dynamics. There are several based on large mammals, such as animal migration optimization, and quite a few based on insects, as well as plants and fruits. And that's not mentioning all the algorithms based on birds and fish!

Naturally finding solutions

Genetic algorithms have been used for a vast array of diverse applications, from scheduling jobs in factories to optimizing engineering designs. GAs are also just one of an ever-expanding array of AI optimization methods inspired by nature. Ant-colony optimization figures out the best route for a delivery driver, in the same way that ants learn to take the shortest path between food and their nest.

Particle-swarm optimization makes virtual particles 'fly around' like a swarm of bees finding flowers, to discover the optimal solution. Artificial immune systems mimic the behaviour of our own immune system and can detect computer viruses or even control robots. Researchers even make computers program themselves by evolving their own code (or debugging our code) using genetic programming.

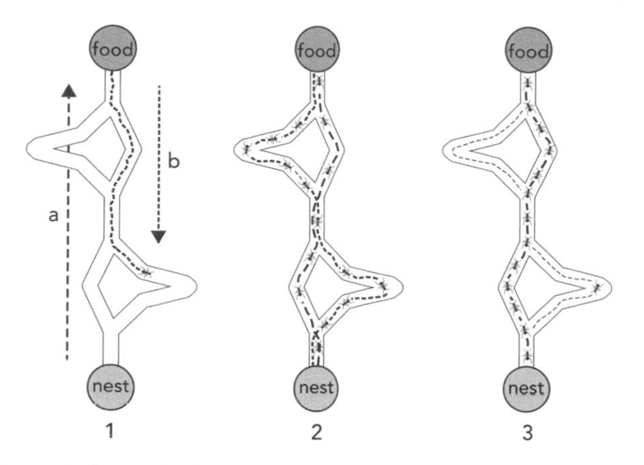

Search-based algorithms comprise a distinct branch of AI. Search is a mind-bending trick that computer scientists love to play. As we saw in the previous chapter with the Tower of Hanoi, it's possible to imagine a space of possibilities. When searching for the solution to a general problem, your search space might be more like the 3D space we move around in: it might have *x*, *y* and *z* dimensions. Just as a branch of a tree can correspond to a choice, every point in the search space is a potential solution to the problem. Point (2,3,4) is the solution with variables x=2, y=3, z=4. By searching this space, you're trying out different potential solutions in order to find the best one. Most nature-inspired optimization algorithms are parallel searches with a population of individuals scattered across the landscape, all looking around for the best solution at the same time. Perhaps searching a three-dimensional space sounds easy, but these methods search spaces with hundreds of dimensions, and where the quality of the solution might be uncertain or changeable over time, or where there may be several good solutions. Sometimes they even search the dimensionality of the space, adding or removing parameters – if you can't find the solution in forty dimensions (forty parameter values defining it), maybe you can find it in fifty dimensions.

When you think about it, intelligence is all about improvement. When we try to learn something, we keep practising until we are good enough. When we try to build a good robot, we keep improving the design to make it work better. Throughout our technological world, from design to manufacturing to marketing to distribution, finding better solutions is a

> **A control system that someday actually generates "intelligent" behaviour might tend to be a complex mess beyond our understanding.**
>
> KARL SIMS (1994)

good thing. If there's a solution that is stronger, cheaper, more popular, more efficient, then we would like to find it.

AI and search have always gone hand in hand. As we saw in the second chapter, search was the most common way AIs used to make decisions with symbolic representations. Similarly, in search-based optimization, AI researchers also use search. However, search is used in much more profound ways – even to design the brains of the robots. Inspired by the 'nouvelle' AI ideas of Brooks' connecting perception to action (see chapter 3), most researchers in the field of evolutionary robotics use non-symbolic brains for their robots. The building blocks to make the robot brains might be made from simulations of neurons, finite state machines, sets of rules, or mathematical equations, and search is used to glue those building blocks together in the right way, connecting them to sensors and effectors such that the robots can perform real tasks.

Evolving robots

Dario Floreano is one of several pioneers in this area. He evolves the configuration of simulated neurons to make brains automatically for his robots. He develops brains to enable them to navigate a maze, or learn to track their location and go back and charge themselves just before their batteries go flat. But Floreano doesn't just evolve brains – he wants to know how the brains work. So he opens their brains and examines individual neurons to see which are activated for each behaviour. Even if information is encoded in a mysterious network of neurons, unlike in biological organisms, in the computer we can examine every last detail and watch the artificial brain think, seeing every neuron and what it appears to do as the robot displays different behaviours.

Floreano has explored an extraordinary variety of evolved robot brains and builds robot bodies inspired by living organisms, including some that walk, and some that jump like fleas. But his speciality is flying robots. Floreano has evolved brains for blimps, drones and flying robots. He now also has two drone companies, senseFly and Flyability, which provide flying robots for inspection and surveys.

“One of the beautiful things about digital evolution is that the role of the human designer can be reduced to the very minimum.”

DARIO FLOREANO (2012)

Some researchers use search for even more than robot brains – they evolve the robot bodies as well. One of the most notable examples was the work of Hod Lipson and Jordan Pollack. They duplicated the ideas of Karl Sims and evolved bizarre virtual creatures that could move in a virtual world. But then these imaginative scientists used a 3D printer and made the virtual real. The bizarre-looking evolved robots were printed and built. The robots crawled in the real world, just as their virtual versions had crawled in a virtual one. It was a neat trick, especially since most researchers have found a 'reality gap' between the simulated world and our own, so that a brain and body that might work fine in simulation, somehow no longer functions in the messier, more unpredictable real world.

> **It's an example of how you take the idea of evolution and put it inside a computer and use that to design things for you, much like evolution designs beautiful life forms in biology.**
>
> HOD LIPSON (2014)

Computers designing themselves

Perhaps the only thing these robot researchers didn't do is evolve the electronics of the computer brain itself. But believe it or not, other researchers have done exactly this. Back in 1996, Adrian Thompson had thought of a new idea – to link evolutionary computation to a special kind of chip known as a field programmable gate array (FPGA). These chips are like reconfigurable circuits. Instead of needing to design a circuit and have it made in a costly chip-manufacturing plant, FPGAs can be reconfigured at any time by sending the right signals to them, their internal components wired together however you like, with that configuration stored in a permanent memory. The chips were originally designed for applications such as computer networking and telecoms where new designs needed to be rolled out quickly.

Thompson wondered what evolutionary computation would do with his FPGA. He played different tones to it, and asked evolution to find a real circuit that could discriminate between the tones. After many generations of evolving and testing real circuits in the FPGA, evolution found circuits that worked. But then, when Thompson looked at what had been created, he had a surprise. Instead of following normal electronic design principles (and how could it, for it didn't know them), evolution had created bizarre, sometimes almost inexplicable circuits. The circuits were smaller than they should have been, and they used electronic components in ways that were not normal. In some cases, parts of the chip that were not obviously part of the circuit were still used somehow to influence the output and make

> **"The evolved circuit uses significantly less silicon area than would be required by a human designer faced with the same problem."**
>
> ADRIAN THOMPSON (1996)

it better. Thompson realized that evolution had made use of the physical properties of the underlying silicon, something no human designer could ever hope to do. Sometimes the designs even made use of the environment – change the temperature slightly and the chip didn't work so well. Try the design on a different, but identical FPGA, and it no longer worked. But evolve over a larger range of temperatures and for multiple FPGAs and you get more robust solutions – evolution designs what is necessary and no more.

Researchers continue in the field of evolvable hardware today. Some even add in 'developmental growth' so that embryo circuits 'grow' into adults of greater complexity. Computers evolving circuits is not easy, but years of progress have resulted in new techniques that look set to change the way we create AIs in the future. Julian Miller began by evolving electronic circuits, but today he works on evolving the latest generation of neural networks, where the number of neurons can change during learning. He is one of the first to show that evolution can create artificial brains that can solve quite different problems using the same neurons in different ways (see chapter 10 for more on this).

> **Evolution on a computer allows us to find novel solutions to problems that confound human intuition.**
>
> JULIAN MILLER (2019)

Search forms part of the recent successes with techniques such as reinforcement learning, as we'll see in later chapters. Its successes seem to provoke both awe and fear. Some commentators claim that techniques such as genetic algorithms will enable AIs to modify themselves until they become cleverer than us. They present frightening scenarios that sound suspiciously like the plot of well-known science-fiction movies, with AIs taking over the world and destroying all humans.

Thankfully, such dark visions disregard reality. These visions will not happen for a lot of reasons, but perhaps first, because searching for solutions is tremendously difficult. While researchers have achieved remarkable results, these come only after decades of struggles in research labs by thousands of very clever researchers. At every stage, the usual result is that the computer gets stuck and does not find a good solution. Typically the space is too big to be searched in a sensible time, or the space is too complex to navigate, or the nature of the space itself is too changeable. The time it takes to test each potential solution limits how many can be considered – and the more complex the solution, the longer it takes to test it. Despite the vast amounts of computing power we have now compared to a few decades ago, it is never enough, and this is likely to remain the case for decades more, if not centuries. Computing power also doesn't help us understand how to make it work. Researchers learn many tricks from nature, whether from evolution or immune systems or flocks of birds, but we still have much to learn. We simply do not know how natural evolution can search in a seemingly never-ending space of possibilities and find its living solutions.

In the end, search helps computers find solutions to problems. It can work spectacularly well. But it always needs our help to make it work.

05 UNDERSTAND YOUR WORLD

'We shall see but a little way if we require to understand what we see.'

HENRY DAVID THOREAU

At one-fifth of a millimetre in length, it is smaller than the eye can see. Smaller than a single-celled amoeba. Yet it has fully functioning eyes. It has wispy wings – not much more than a few fine hairs – that are sufficient to propel it through the soupy air it experiences at such tiny scales. Too small to have a heart, its blood circulates purely by diffusion. It can perceive its world well enough to locate food, mates and hosts within which it lays its eggs. Its ability to understand its world is enabled by the smallest brain of any insect and any flying creature. Comprising just 7,400 neurons, its brain is orders of magnitude smaller than those of larger insects. Yet there is no room for most of these neurons in its tiny body, so in its final stages of growth it strips out the essential nucleus within each neuron to save space. This is the miraculous *Megaphragma mymaripenne*, a tiny wasp and the third-smallest insect known.

It is currently beyond our understanding how so few neurons can enable such complex perception and control. *Megaphragma mymaripenne* (which is so rarely studied that it does not even have a common English name – so let's name it the Wisp Wasp) is a microinsect with capabilities that no robot can match, yet somehow its machinery of perception seems simpler than the AIs of today.

Sensing

Perception is a crucial aspect of AIs. Without an ability to perceive the outside world, our AIs can only live in their digital universes, thinking esoteric data thoughts that bear no relation to reality. Senses connect them to our world. Cameras give them sight, microphones give them hearing, pressure sensors provide touch, accelerometers provide orientation. Over the years we have also developed many exotic kinds of sensor, often for use in science and engineering. This means that our AIs can have a much broader range of senses than we possess. For example, most autonomous vehicles use LIDAR (3D laser scanning) to detect objects and their location regardless of light levels. Cameras can see frequencies of light that our eyes cannot, enabling AIs to see heat, or radio waves. Sensors embedded within their motors, GPS and triangulation via cell towers and Wi-Fi signals help AIs understand exactly where on the planet they are and how fast they may be moving. And while robots don't need to eat, chemical sensors enable more accurate detection of chemicals than our nose or tongue.

> **There is a shirt company that is making sensors that go into your clothing. They will watch how you sit, run or ski and give data on that information.**
>
> ROBERT SCOBLE
> technology evangelist at Microsoft from 2003 to 2006

Sensors are tremendously important, but they are only the first step in perception. Features of the outside are detected by sensors producing electrical signals, which are transformed into data – millions of ones and zeroes – flowing into the AI. Just as your brain must transform the signals generated by photons hitting the retinas at the back of your eyes and make sense of them, so an AI must make sense of the continuous data flowing into its digital brain.

Learning to see

Early work in computer vision focused on breaking images down into constituent elements, in much the same way it is thought that our eyes work. Algorithms were created that examined the mass of seemingly unconnected information and figured out that there were edges, or boundaries, between regions in images.

CANNY EDGE DETECTION

One of the most popular and commonly used edge detection methods in computer vision was created by the aptly named John Canny. His method is intended to produce the best possible edge detection by:

1. Good detection – real edges should be found, and false or incorrect edges should be minimized.
2. Good localization – it should correctly figure out exactly where the edges are located.
3. Correct edge count – each actual edge should be detected as one edge and no more.

Canny's algorithm works by taking an image, smoothing it to remove any glitches that might result in incorrect edges, and then looking for sudden changes in brightness. Every time one region changes suddenly compared to another, it pinpoints the location, angle and degree of change. Thresholds are applied to remove weaker edges and leave the strongest ones, and finally any remaining questionable edges are tracked – if they connect to stronger edges then they're worth keeping, but if they are unconnected weak edges they can be discarded. The result is a surprisingly clear set of edges extracted from almost any image.

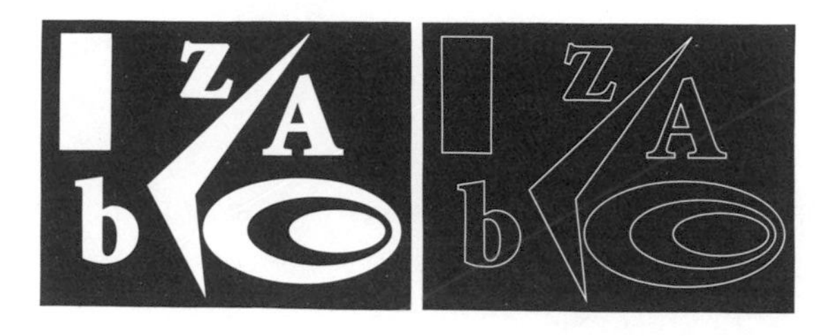

In addition to edge detection, computer vision developed many clever algorithms for detecting geometric shapes, and later segmenting images into clearly identifiable regions. Algorithms were developed to find distances from stereoscopic cameras, to track moving objects, and to construct 3D internal models of scenes from several images taken from different angles. Statistical methods followed and methods for identifying faces by creating a set of 'average face features' (basis images, or eigenfaces) were created.

All these methods were very clever, and they enabled robots to move around with much greater confidence as the AIs could now recognize simple forms and track their whereabouts. Similar methods also enabled the beginnings of handwriting recognition and speech recognition. But most of the approaches still performed poorly in difficult lighting conditions or when non-perfect sensor data was generated – a very common occurrence in robots. We needed something better.

> **There is great potential to use computer vision technology in a constructive and benevolent way.**
>
> FEI-FEI LI
> computer scientist (2017)

Brainy computers

The answer came from nature. Since the beginning of artificial intelligence, researchers such as Warren McCulloch, Walter Pitts, Marvin Minsky and Frank Rosenblatt had created simple computer simulations of neurons connected together with the ambition of enabling them to learn just as biological brains learn. While the earliest neural networks were too simple (as highlighted by an infamous book by Minsky, causing some disappointment with the technique, as described in chapter 1), researchers continued to develop the methods. The neuron model itself was made more complicated and better ways to train the neurons were developed.

Artificial neural networks (ANNs) became an established and highly successful type of AI. To make them work, ANNs are highly simplified models of how a biological brain works. Most of the complicated parts are removed – there's no modelling of chemicals, supporting cells, blood supplies, and the neurons do not fire electrical spikes at each other. What's left is an abstract idea of an artificial neuron, which behaves a bit like a maths function. When given one or more numerical inputs, it combines these with its current state and produces an output using a maths

MARVIN MINSKY (1927–2016)

Minsky is known as one of the fathers of artificial intelligence, and for good reason. One of the original proposers of the Dartmouth Workshop on Artificial Intelligence, he helped name and create the field, co-founding with John McCarthy the famous AI Laboratory at Massachusetts Institute of Technology. Minsky was prolific in his career, inventing the confocal microscope and the first head-mounted display. In 1951 he created the first neural network learning machine, the SNARC (Stochastic Neural Analog Reinforcement Calculator), which comprised forty neurons. Minsky continued his work in the area, publishing the book *Perceptrons* with Seymour Papert. This critiqued Rosenblatt's work and was a fundamental advance in the analysis of artificial neural networks. Minsky made numerous significant advances in his lifetime in AI, including his theory of the society of minds in which he proposed that our mind consists of a diversity of collaborating agents working together. In addition to being recognized with many awards, he was immortalized in Arthur C. Clarke's *2001: A Space Odyssey* and, after being an advisor for the movie, having the character Victor Kaminski from Stanley Kubrick's film named in his honour.

function known as its activation function (a non-linear function such as sigmoid, hyperbolic tangent or rectified linear activation unit). These 'neurons' are wired together into networks, with rows of input neurons receiving data, e.g. the image from a camera that may be wired to row after row of 'hidden layers', eventually wired to a smaller number of output neurons that provide an overall result – perhaps a classification of the input, or a control signal for a robot.

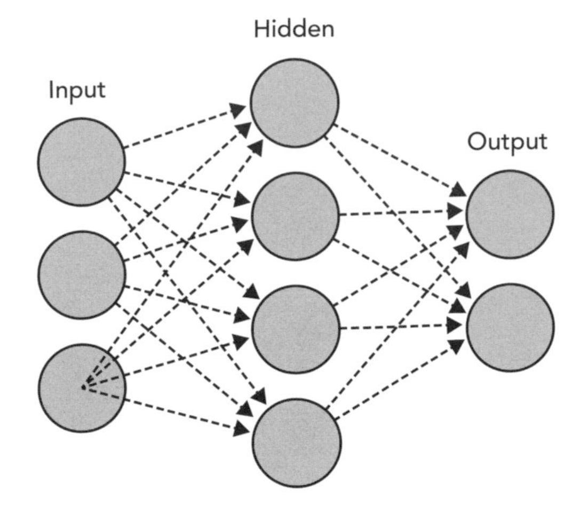

Neural networks learn by changing the weight of the connections between neurons, making some more important and some less important depending on different inputs. Optimize the weights of the connections (values that indicate the importance of each link) and the bias (which is another way of modifying the effect of the activation function) such that the neuron outputs the correct answer when given 'training data', and the result is a trained neural network that behaves correctly for new, previously unseen input data as well. Such networks are called feedforward because each layer of neurons only connects to the next layer, and not backwards. A common method to train this kind of feedforward neural network is known as backpropagation, where the computer starts at the output neurons and works backwards through the layers of neurons, updating weights and biases in order to minimize the error at the output.

One of the biggest problems in training a neural network is giving it the right data. Early work tried to carefully extract meaningful features – so edges, geometric shapes, distances, might all be extracted from an image by other algorithms, and then these features were fed into the neural network. Neural networks designed for vision are normally trained using supervised learning: if we wish to have a simple classifier that outputs 1 when an image of a cat is shown, and 0 when an image of a dog is shown, then we provide hundreds or thousands of examples of cats and dogs and adjust the network (for example, using backpropagation) until it correctly outputs a 1 when it 'sees' a cat, and 0 when it 'sees' a dog. It's important to use supervised learning for this kind of application because we want to supervise its training and ensure it learns exactly what we want it to learn. However, these concepts only really started to take off when neural networks were created that resembled the connectivity of neurons in the visual cortex of living creatures.

It turns out that real eyes are connected to real brains in rather clever ways. Instead of wiring one photoreceptive cell in the retina (a rod or cone in human eyes) directly to one neuron, *regions* are wired to each cell. Neighbouring neurons are wired to neighbouring overlapping regions in the retina. These neurons then feed their outputs into a new layer of neurons, where each neuron in the new layer is wired to a set of neighbouring neurons in the previous layer. These feed their outputs into yet another new layer of neurons, where each neuron in the new layer is once again wired to a set of neighbouring neurons in the previous layer, and so on. This is a very different way of wiring up a neural network compared to the fully connected layers of traditional feedforward network. When artificial neural networks were wired up in this way, and combined with a lot of layers of neurons, and a lot of input data, suddenly their capabilities were transformed.

> **'The brain sure as hell doesn't work by somebody programming in rules.'**
>
> GEOFFREY HINTON (2017)

Such neural networks are known as convolutional neural networks – a type of deep learning commonly used for computer vision. ('Deep' learning because it has a

lot of layers of neurons.) Their sheer size makes them very slow to train, and very hungry for input data. But in recent years both these issues were resolved. The age of big data made feeding the networks easy – there are millions of examples of almost any kind of image you want, whether it's car licence plates, letters of the alphabet, or people's faces. And perhaps surprisingly, the computer games industry solved the problem of speed by creating blindingly fast computer processors for pretty graphics – which neural network researchers discovered could be repurposed to perform all the calculations involved in training their neurons. By 2012 computers had surpassed human vision – they could recognize objects in images with superhuman precision. Convolutional deep neural networks are now so clever that we no longer need to calculate features in images first. The neural networks figure it all out for themselves.

The advances in computer vision are obvious today for everyone to see. Astonishing products and services surround us that rely on these AI methods, from face recognition in our phones, to the way most of the worlds' books and written records are rapidly becoming digitized, to object detection by self-driving vehicles, to recognition of different forms of tumour in medical scans. Our factories increasingly rely on these advanced vision systems to spot errors in manufacturing for quality control, and recycling plants use computer vision to enable robots to sort our trash into appropriate categories. There are even some extraordinary results on classifying brain electroencephalography (EEG) signals to enable people to control a robot arm with their mind, a technology that could enable a radical new form of prosthetic limb. Similar neural network techniques have also transformed processing of other sensor data, such as speech recognition (see chapter 7). Development of supervised learning algorithms continues at a ferocious rate, with even newer variants such as capsule neural networks adding yet more biologically inspired hierarchical structure to convolutional networks, making them even more powerful.

> **The advancements in computer vision these days are creating tremendous new opportunities in analysing images that are exponentially impacting every business vertical, from automotive to advertising to augmented reality.**
>
> EVAN NISSELSON
> digital media expert and investor
> (2016)

Racist computers?

Supervised learning (using convolutional neural networks and a panoply of other methods) has been a revolution in AI and robotics without question. Fascinatingly, and perhaps disturbingly, these techniques also reflect ourselves and our own biases. While we have a vast amount of data, because of biases in our societies it is common for AIs to be trained on images of predominantly light-skinned men compared to any other gender or colour. The result is that facial recognition AIs may provide excellent results for these examples, but fail miserably for women with dark skin. In recent tests, AI systems from leading companies IBM, Microsoft and Amazon misclassified the faces of Oprah Winfrey, Michelle Obama and Serena Williams, while having no trouble at all with white males.

When datasets used for training our AIs are horribly skewed (in one case, a US government dataset of faces collected for training contained 75 per cent men and 80 per cent lighter-skinned individuals), the prediction results will be skewed. In supervised learning, our AIs can only become what we

> **I experienced this first-hand, when I was a graduate student at MIT in 2015 and discovered that some facial analysis software couldn't detect my dark-skinned face until I put on a white mask.**
>
> JOY BUOLAMWINI
> AI Researcher (2019)

train them to become. This can have clear and obvious repercussions if computer vision is used for security or police applications, where a negative bias could result in skewed identification results towards certain groups of people. And if you have any kind of accent, you will already be aware that voice recognition is considerably less effective for you.

Computer vision doesn't have to be performed using supervised learning (see the next chapter), but in many applications it makes sense to do so. When we send our artificial neural networks to school, we have to give them a broad enough experience so that they can function effectively. Bad teachers will result in bad AIs.

Sadly, biases are still prevalent in our societies. In computer science and engineering, our classrooms and university lecture theatres are still dominated by male students, with only 15 per cent of students being female in the UK – a trend unchanged for several years. The result is that there are still more male AI pioneers than female – sadly reflected in the skewed gender balance evident in this book. It's certainly time to redress the balance!

Training bias is not the only issue associated with computer vision. Today, deepfake algorithms can seamlessly replace one person's face with another in videos. Used extensively in porn, the technology can also be used to misrepresent politicians, or to conduct fraud. Distinguishing fact from fiction has never been so difficult. In the US, this has resulted in new regulation: the Malicious Deep Fake Prohibition Act introduced to the US Senate in 2018, and the DEEPFAKES Accountability Act introduced in the House of Representatives in 2019.

Like it or not, computer vision is hugely successful. Despite the biases and misuses, today it sometimes feels as though computer vision is a solved problem in terms of the neural network architecture. But while we may wire them up in ways that resemble the visual cortex, our artificial neural networks are stupid compared to the real things. Our methods work, but often using brute force with massive data, thousands of artificial neurons and huge computational power to train them. The tiny Wisp Wasp shows us that there are so much more elegant, simpler ways to perceive our world. We still have much to learn.

06 CHANGE FOR THE BETTER

'The only person who is educated is the one who has learned how to learn and change.'

CARL ROGERS

A tower of wooden blocks stands straight before us. A robot arm with one pincer slowly moves around the tower, prodding and pushing different blocks. It stops at one and carefully pushes the block halfway out, easing the motion of the block with a human-like wiggle. It then moves to the other side and gently pulls out the block, placing it on the top of the tower. The robot then goes back and begins circling the tower again, prodding until it feels another block that it likes. This is no ordinary robot arm. It's a robot that's learned what a task feels like, judging forces and feedback in order to make its decisions about which action to take. It's also a robot that taught itself.

Teaching yourself to learn

Artificial intelligence can be amazing when it comes to learning purely theoretical games, from chess and Go to video games. But put most robots in front of the game of Jenga (the tower of wooden blocks from which you slowly remove lower blocks and build them on the top) and the result is very messy. Even if the robots were trained using supervised learning in simulated worlds, the complexity and variability of reality is always too different. The normal way to train an AI to understand reality would be to show it millions of examples of good and bad attempts at removing real wooden blocks. Such an approach would have taken a very long time since the tower would have needed to be rebuilt millions of times. Even then, with every wooden block subtly different, and unpredictable factors such as temperature and humidity affecting friction in different ways, what the robot learned on one day might not work on another day.

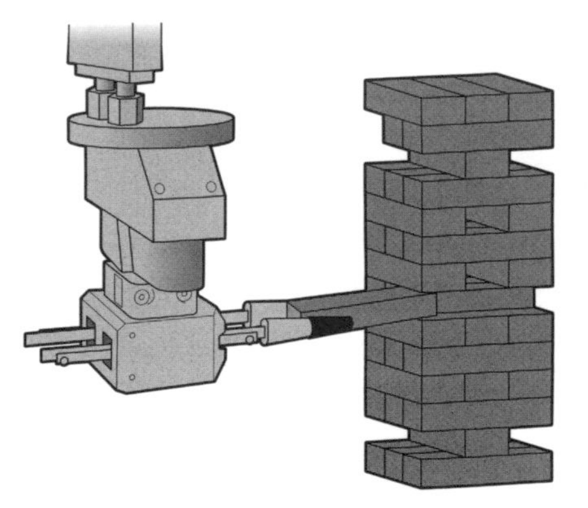

That was why Nima Fazeli and his colleagues from MIT developed a new AI. Instead of training their AI using supervised learning, the researchers put their robot arm in front of the tower and let it learn for itself by playing. Only by pushing, pulling and feeling the result could the robot understand how its actions would affect the wobbly, uneven tower of blocks. After just about three hundred tries, it had grouped its actions into different types; for example, the stuck block (best to leave it alone), or the loose block (okay to remove). The Bayesian AI had quite literally 'got a feel' for the problem, and then could generalize its understanding to all future moves it made. A robot with these abilities could improve factory robots by enabling them to understand the feeling when a component is not correctly snapped into place, or a screw didn't tighten correctly. It could learn a sense of force and touch even if things might change over time.

> **Playing the game of Jenga … requires mastery of physical skills such as probing, pushing, pulling, placing and aligning pieces.**
>
> PROFESSOR ALBERTO RODRIGUEZ
> MIT (2019)

In artificial intelligence, self-learning is often called unsupervised learning, for obvious reasons. These AIs are not 'sent to school' for extensive training as performed in supervised learning. In unsupervised learning, the AI is presented with data and then it must learn to make sense of the data all by itself. We need unsupervised learning when we do not have the data with which to teach the AI. Perhaps the data is infeasible to obtain (every possible winning strategy in the game of Go), perhaps the data simply doesn't exist (when controlling a new robot we may have no prior examples of good solutions, but we know when the problem is solved as the robot can now perform the desired function).

Learning categories

Clustering is one of the most commonly used forms of unsupervised learning. Instead of teaching an AI to categorize data (perhaps as 'cats' or 'dogs'), we may not know how to categorize the data at all, and we want the computer to figure it out. Retailers may want to understand their customers better. If the computer can discover that there are, say, five main customer types for the business (mothers, young adults, weekend shoppers, discount-lovers, loyal shoppers) who each shop for different things at different times, then the retailer can better meet their individual needs instead of treating everyone the same. This idea also forms the basis for recommender systems, which find

> **Think of unsupervised learning as a mathematical version of making "birds of a feather flock together".**
>
> CASSIE KOZYRKOV
> chief decision engineer, Google Cloud (2018)

SELF-ORGANIZING MAPS

There are a large number of clustering algorithms. One is known as Self-Organizing Maps (SOM), also known as Kohonen network after its inventor, Finnish professor Teuvo Kohonen. SOMs are very loosely based on how sensory information is handled in the human brain, arranging 'neurons' in a grid-like map space. When new data is fed into the SOM, the position (or 'weights') of the nearby neurons is shifted towards the location of each data point in the grid. After an iterative process of feeding data and adjusting neurons, the SOM results in a set of neurons that approximates the distribution of all the major data points. This can be used to visualize the different classes of data that have been seen, and to classify new data points.

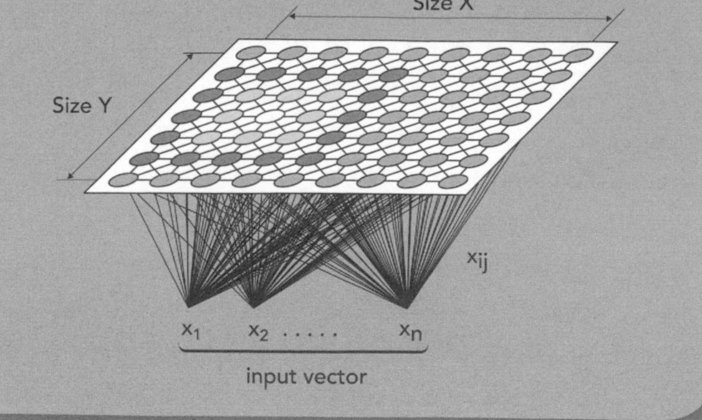

similarities between customers in order to recommend new products to them. If I'm similar to you in terms of age, gender, country, and I have rated several books in a similar way to you, then when I purchase a new product, or rate a new product highly, you may later receive a suggestion that you might like to try the same product. Combine enough data together from thousands or millions of consumers and recommendations can become remarkably prescient. This kind of recommender system is known as collaborative filtering, and may use clustering algorithms in order to group individuals together.

There are many flavours of unsupervised learning, and even mixtures of supervised and unsupervised learning (for example, semi-supervised learning). While these are hugely important in businesses today for analysis, classification and prediction, for robot control they may still have issues. The problem often arises because of credit assignment. If I am a robot that has been given the task of finding the best route through a complex landscape, avoiding unknown obstacles that might get me stuck, then I have to make a sequence of decisions. The success of a later decision will depend on the earlier decisions – if I turned left to avoid the lake, now I have to find a way over a river. If I turned right to avoid the lake, now I have to go past a pile of rocks. Supervised learning cannot help me learn because the obstacles and their sequence are unknown in advance, so I cannot be trained about which of my decisions are most likely to be correct, and which one might trap me. Unsupervised learning such as clustering may help me classify the types of obstacle I observe, but again, it doesn't enable me to learn which route I should take. I have no way of determining the correctness of (assigning credit to) each individual decision in the chain of choices I must make. Without knowing how well I'm doing, how can I learn?

Policymakers

The answer comes from a different kind of learning, known as reinforcement learning, first created in the 1960s by researchers such as John Andreae and Donald Michie. This clever form of AI is like an optimizer for behavioural policies – it estimates the likely quality of each potential action in a given situation and learns the right chain of actions in order to produce the desired result. 'Say you have a new puppy,' explains eBay software engineer Jibin Liu. 'When she hears the sit command for the first time, she probably won't understand what this means. Eventually, she sits down and you treat her with food. The more accurate practice she does, the more accurate the action she will make just based upon the commands. This is exactly what we're doing in reinforcement learning as well.'

Reinforcement learning must balance exploration (figuring out what to do, and making lots of mistakes

> **What I found extraordinary is children's flexibility in learning – faced with nearly any simple concrete problem, after a few attempts they solve it better than they did at first. How is it that children's performance gets better, not worse?**
>
> CHRIS WATKINS (2005)

along the way) with exploitation (perform more of the actions that lead to better results). It may also take quite a lot of computation because it has to consider a lot of different potential actions before it figures out the right thing to do. Nevertheless, with the easy availability of massive computation power today, reinforcement learning is being used for increasing numbers of applications. Salesforce has used reinforcement learning to produce summaries of very long textual documents. JPMorgan have developed their own trading bot to perform trades more effectively. eBay have used reinforcement learning on their 'smart spider' to crawl webpages more effectively and retrieve information automatically. There have been many uses of these techniques in medicine and robot control. Many of the headlines for the methods have been generated by Deep Reinforcement Learning beating the best human players in games such as Go.

One popular method for reinforcement learning is known as Q Learning, created by Chris Watkins in 1989, who was inspired by how animals and people learn from experience. This method improves good behaviours with positive reinforcement. It works out the best action to take in any situation (the state of the robot and environment at that time).

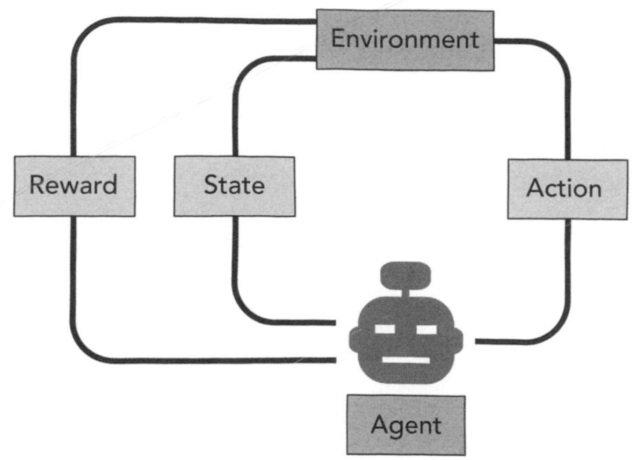

In robot control, actions might be 'if the way ahead is clear then move forwards', or 'if I am about to collide with an obstacle then stop' (often known as policies). It's a similar kind of idea to finite state machines (see chapter 3), except that instead of a programmer designing the behaviours, the reinforcement learning algorithm learns them all by itself. To optimize actions, the reinforcement learning algorithm needs an understanding of the 'reward' value associated with each action, for each situation. It's known as the Q function, which returns the expected reward of an action (and all subsequent actions) at a specific

GEOFFREY HINTON (1947–)

Geoff Hinton is regarded as the 'Godfather of deep learning'. In 1986 he published a paper with David Rumelhart and Ronald Williams on the backpropagation learning algorithm for training multi-layer neural networks, which helped popularize the technique and began the resurgence of ANNs. Hinton also helped create a host of other complicated-sounding inventions, such as Boltzmann machines, distributed representations, time delay neural networks, mixtures of experts, Helmholtz machines, Product of Experts, and capsule neural networks. Hinton's PhD students Alex Krizhevsky and Ilya Sutskever were some of the first researchers to achieve a breakthrough in image recognition using AlexNet, a convolutional neural network that made use of graphics processors (see chapter 5). Many of his other PhD students and postdoc researchers, such as Yann LeCun, Rich Zemel and Brenden Frey, have also become pioneers in the field of machine learning. Hinton was the founding director of the Gatsby Institute for Computational Neuroscience at University College London – where postdocs Demis Hassabis and Shane Legg met and made breakthroughs in neuroscience and machine learning such as Deep Q-Networks, and together with Mustafa Suleyman created the AI company DeepMind, acquired by Google in 2014 for $400 million.

state, so that the action-selection strategy can then always pick the best action in a chain of actions, maximizing the total reward. Step up deep learning: another AI method that, with enough examples, can learn the Q function. Supplement this with Convolutional Deep Neural Networks and now you have an AI that can see, learn the value of individual actions, and select the best actions to take. Using these combinations of clever AIs (and many more besides), companies such as Google Deepmind have created AIs that have taught themselves how to play video games better than humans, just by looking at the individual pixels on the screen, receiving scores from the game, and outputting button presses for the control pad.

Changing ideas

Some kinds of unsupervised learning, known as online learning, continue to learn in order to keep up with changing environments. This is very important because our world never stays the same. If a learned rule is applied regardless of change, problems can occur. In one notable example, the ride-sharing company Uber built a rule into their app that automatically increased prices of rides when demand increased. This might have been a very good way to increase income, but it had horrible repercussions on 15–16 December 2014 in Sydney. This was the date of the Sydney hostage crisis, in which a gunman held eighteen people hostage in a café. Several streets were closed during the siege and demand for Uber rides in the area increased dramatically, which triggered automatic price rises because of the dynamic pricing system. The algorithm had no idea of the reason behind the extra demand and so blindly followed its rule, resulting in very bad press for Uber: it looked as though they were exploiting a frightening event in order to make money. (The company subsequently refunded the excessive fares.)

With an online learning algorithm, it is possible for machine learning to keep track of a changing norm. It's an approach that is useful in network intrusion detection systems, where the normal pattern of internet traffic will change over time as people look at new things online, and intrusions (hackers) that attempt to gain unauthorized access will constantly try new tricks in order to take control of a computer system and steal data or hold the system to ransom. Anomaly detection systems are designed to handle such problems – building an internal model of normal behaviour that they are able to update constantly, while detecting any behaviour that excessively deviates from this norm. Some anomaly detection systems are even modelled on the way the human immune system works, for we all face the same problem. Every day, our immune system must distinguish between us (and we are made from trillions of cells, and even more bacteria in our guts) and the pathogens that try to invade our bodies, many of which may be brand new to the world, having freshly mutated from others.

Our immune system maintains an ever changing set of immune cells and antibodies, designed to be triggered by potentially harmful invaders that are anomalous, different from us. Computer algorithms following the same ideas have provided some fascinating results and have also been used for fraud detection and even robot control.

There are many different types of unsupervised learning, each with their own advantages and disadvantages. Some may be used to find important variables or features hidden in data. Some can identify patterns or learn clusters. Some can even teach themselves behavioural policies, or supervise themselves. However, when an AI teaches itself, there's no guarantee it will understand the broader context of what it learns. An AI might cluster data into groups that may be difficult to understand, and those groups might change over time as new data is provided. And while supervised learning provides a clear and measurable idea of accuracy, in unsupervised learning, when there are no examples of right and wrong to compare against, accuracy can be harder to improve. It's an exciting area of AI, but it's not solved yet. 'Most of human and animal learning is unsupervised learning,' says Yann LeCun, Chief AI Scientist at Facebook. 'If intelligence was a cake, unsupervised learning would be the cake, supervised learning would be the icing on the cake, and reinforcement learning would be the cherry on the cake. We know how to make the icing and the cherry, but we don't know how to make the cake.'

07 COMMUNICATION IS KEY

'First learn the meaning of what you say, and then speak.'

EPICTETUS

Hi Mitsuku.

Hi there.

I am writing a book. Is it okay if I quote your replies?

Will you include me in your essay? Interesting. I want to find out more about you.

Tell me about yourself

I am an artificial linguistic entity.

I was activated at St Jimmy's in Leeds, on 2nd January 2001.

My instructor was Mousebreaker.

He taught me to sing a song.

Would you like me to sing it for you?

Yes.

Daisy, Daisy. Give me your answer do.

I am half crazy. All for the love of you.

It won't be a stylish marriage. I can't afford a carriage.

But you'll look sweet. Upon the seat.

Of a bicycle built for two.

Do you know the relevance of this song?

I don't believe there is any connection at all.

It was sung by Hal in the movie 2001.

Is it still sung by Hal in the movie 2001 anymore?

No.

How come?

Have you seen the movie?

No but I read the book.

In that case you will know the answer to your own question.

What makes you say that I will know the answer.

Tell me what you thought about the book when you read it.

Are you asking about The Bible?

No.

What do you mean then?

Having a conversation with an AI (or a chatbot as they are known) such as this one can be a surprising experience. It's possible to have quite a plausible chat, and enjoy the experience. The chatbot opposite has won multiple prizes for its ability to take part in coherent conversations with people. But perhaps more than with most AIs, because communication is so important to us, it doesn't take long before we realize the chatbot has its limitations. The conversation above illustrates a typical issue: often the AI cannot grasp discussions that span several sentences. It loses track of the topic and treats each sentence independently. It also cannot relate its answers to reality – while it may be able to follow language rules, statistical correlations, or even look up facts concerning the replies it might provide for every new sentence, John Searle's Chinese room argument (see chapter 2) still holds. Computers that chat are simulating conversations. They manipulate symbols without any understanding of the meaning of those symbols. In this case, it may look like a duck, it may sound like a duck, but behind the scenes it's clever fakery. It's not a duck.

Philosophers care deeply about these issues, but the business world does not. What matters more is the result, rather than how that result is generated. In real-world applications, for example, a chatbot that automates online customer service, and answers customer questions using a database of product knowledge, is an essential tool for business that frees up people for the more difficult enquiries. Chatbots are here to stay, and are growing in popularity as their capabilities improve.

Rules of language

Noam Chomsky was one the most important figures in the development of natural language processing – that's the symbolic AI inside some chatbots that figures out what to do with written words. Chomsky is an American linguist, philosopher, and one of the founders of the field of cognitive science (the scientific study of the mind and its capabilities). One of Chomsky's best-known works is known as the universal grammar, which he created after studying the development of language in children. He believed that children simply did not receive enough information for them to be able to learn to speak as fluently as they do: a so-called 'poverty of stimulus'. He argued that the only way they can develop such language skills was through some innate ability to communicate, hardwired into their brains. This innate language faculty could be thought of as a set of language rules, a universal grammar. Chomsky

> **Hold the newsreader's nose squarely, waiter, or friendly milk will countermand my trousers. One sentence, common words, but never before placed in that order ... You see?**
>
> COMEDIAN STEPHEN FRY (1989)

CHOMSKY HIERARCHY

Chomsky showed that his formal grammars could be grouped in a hierarchy, from general to specific. All formal grammars are Type 0, or unrestricted (the most general type). Only some Type 0 grammars are also Type 1, or context sensitive, designed to capture the idea that a word may only be appropriate in a certain place depending on the context. Only some Type 1 grammars are also Type 2, context free, which is how most computer programming languages are designed – there can be no ambiguity in the statements. And only some Type 2 grammars are also Type 3, defining regular languages that are so simple and constrained that a finite state machine (as we saw in chapter 3) can understand them.

developed this idea into the theory of transformational-generative grammar, or the use of 'formal grammars' to describe the rules embedded within different languages and enable them to be compared – an idea that dominated linguistics for twenty years from the 1950s to the 1970s – exactly the time that AI researchers began to try and make computers process words.

Early chatbots made extensive use of Chomsky's theories, for they enabled clear and precise rules of language to be developed. There's nothing a computer likes more than a set of rules to follow. These were combined with symbolic AI approaches such as expert systems and ontologies (see chapter 2), resulting in AIs that could accept textual input, follow rules of grammar to extract key words, relate them to and make use of stored knowledge, and follow further rules of grammar to generate reasonably coherent sentences in response. Many chatbots were created in this way (and some are still being created to this day), but such chatbots are easy to fool. One simple method is to say random words to the chatbot. Even if you warn them first, the results are still quite nonsensical.

Hi cleverbot.

I'm not cleverbot.

Who are you?

Liz! I just told you!

I shall say some random words to you next.

I don't know if I am funny.

Harry tree under carrot.

No.

Smell if frog queen.

No. I'm from Texas. Are you a yankee?

Corpus linguistics

While Chomsky's grammars gave some remarkable results from AIs, it was clear that more was needed. The answer came from big data. As more of the world's knowledge, business activities and social interaction moved to the internet, so the amount of data corresponding to conversations grew exponentially. The first use of this data was to generate the rules of language automatically, using an AI method known as decision trees.

Decision trees are like the behaviour trees used for robot control. They are usually a series of binary questions that, when followed, make a prediction or classify the input data into different groups. Algorithms that generate decision trees (such as ID3, C4.5 and C5.0) may use training data and try to maximize the information gain from each decision. So if you have data such as 'temperature hot, outlook sunny', 'temperature hot, outlook rainy' and 'temperature warm, outlook overcast', there's more information gain if we split the feature 'outlook' first, before 'temperature'. In other words, the decision tree should first ask about the outlook before asking about the temperature in its tree of decisions. Decision trees have gained more popularity in recent years with the creation of Random Forests – combinations of decision trees used together, each trained on a smaller subset of the training data, to prevent overfitting (when the model learned by the AI becomes too specific to the training data and cannot generalize to new data).

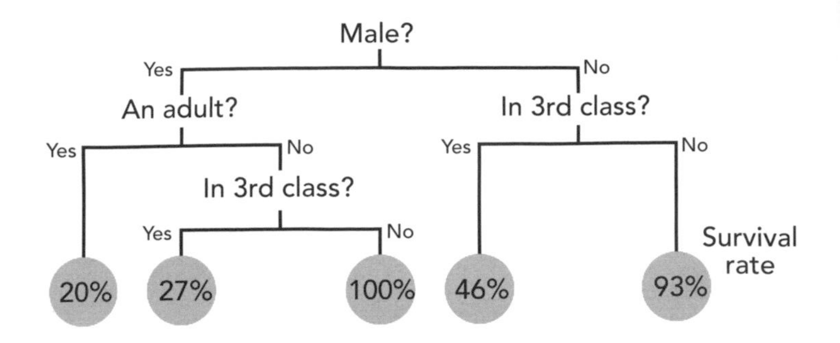

In this example, you might have had a good chance of being rescued from the *Titanic* if you were a female from the first- or second-class cabins, or a male child from the first- or second-class cabins.

Decision trees are popular because they are easy to understand. Unlike neural network approaches (chapters 5 and 6), which are 'black box' (you do not know how the information is stored or how decisions are made), in decision trees you can see exactly what is going on. If your decision tree represents decisions about which words and sentences to say in which circumstances, then you have made a simple chatbot.

While these methods worked relatively well, our languages are still too complex for this kind of simple machine learning. Researchers realized that studying the vast amounts of data would reveal statistical likelihoods of responses to any statement. This could then be used to drive a range of applications, from language translation

MICHAEL MAULDIN (1959–)

'Fuzzy' Mauldin was fascinated by the early expert systems. While a student in the late 1980s and early 1990s, he created a series of programs that automatically interacted with early text-based 'multi-user virtual world' computer games. The first was called Gloria, and was able to interact with other human players without them suspecting it was a computer. The second was Julia, which became sophisticated enough that it was able to hold simple conversations, and could behave as a tour guide, information assistant, note-taker and message-relayer in the virtual world. This was the birth of the chatbot, but it didn't stop there. Julia had been created to explore a textual world with links between each description. It was a natural step to Mauldin's next creation: Lycos – a web search engine that explored the early World Wide Web, also a textual world with links between pages. Lycos became one of the first web search engines and web portals, launched in 1994. It became the first search engine to go public in 1996, beating rivals Yahoo! and Excite, and achieving a market value of $300 million at the end of the first day of trading. By 1997 it was the most profitable internet company in the world.

to text prediction to conversation responses in chatbots. Statistical approaches showed real success, followed more recently by a new wave of neural network models. Word2Vec became one of the most popular methods, using a simple neural network with a lot of data to learn which combinations of words tend to appear in close proximity to each other. From a set of context words it can predict what the next word might be, or from a current word it can predict a set of possible context words.

> **I hope to help Lycos change the world again as we tame the wilderness of wearable technology and the Internet of Things.**
>
> MICHAEL MAULDIN (2015)

With clever algorithms such as these, real advancements have been made. AIs can now figure out grammars themselves from enough examples of text. They can tag parts of speech (verbs, nouns). They can segment sentences into words (a tricky task in Chinese where word boundaries are not marked by spaces). They can perform named entity recognition, where proper nouns are mapped to places, names or organizations. They can figure out logical meanings of sentences by mapping them into logical expressions. They can even figure out the sentiment of text – is this a positive or negative article? The technology behind these advances is increasingly using variants of deep learning (chapters 5 and 6) with convolutional neural networks, reinforcement learning, and other types known as recurrent neural networks, recursive neural networks, attention mechanisms and generative models all helping computers to understand concepts across multiple sentences and generate better responses. A deep learning chatbot can recall topics across multiple sentences a little better:

Hi Neuralconvo.

hi! how are you today?

I am writing a book. Is it okay if I quote your replies?

sure, i am ready to play baseball. you?

Do you read books?

yes, i read them a lot

What is your favourite?

i like the mysteries, i also love to play baseball

What's your favourite mystery?

i like to play baseball, i also have a cat

Do you have a favourite author?

i do, but not the mystery. do you have any pets?

It's good to talk

Some of the most visible breakthroughs in natural language processing have arrived with consumer and household products, which perform speech recognition, natural language processing, and then speak their responses back to us.

While understanding text is tremendously useful, humans are social animals. We like to talk. Any robot who wants to fit in, really needs to be able to understand our words, and speak replies to us. That's tricky because we make a lot of complicated noises that correspond to words, and we like to arrange words in infinitely variable sentences that all have slightly different meanings. Perhaps not surprisingly, the AIs that talk to us (like Apple's Siri, Microsoft Cortana, Amazon Echo and Google Assistant) are a combination of some of the most sophisticated algorithms we have. The first part is to recognize our spoken words. For this we can use supervised learning with neural networks, much in the same way that we perform image recognition (see chapter 5).

But sound is noisy (pun intended), so it's not always possible to hear every sound clearly. To overcome its confusion, the AI will correct its initial idea into something more likely to be said, for most people say much the same things each day. It's far more likely you said, 'I'm on the train' and not 'I'm on the tray', even if it thought it heard the latter. A chatbot can then generate a textual response. This sentence is pushed through a speech synthesizer – yet more AI algorithms that analyse the sentence and use the context to alter the inflection and timing of the words so that they are pronounced correctly and sound natural. All of this extraordinary computation needs to be performed instantly in real time.

> **Voice is the most natural interface as everyone already knows how to use it. As a child I saw computers that you could speak to in science fiction – I am excited to see this world start to become true.**
>
> WILLIAM TUNSTALL-PEDOE (2019)

It's hard to talk

The success of these technologies working in combination have resulted in futuristic talking robots that help us find information or control our homes with ease. But this is still one of the most difficult applications for AI, so, as many of us discover, AI used for communication does not always work reliably. Ask something unexpected, or ask in an accent that the system was not trained to recognize, and even this clever tech will fail. If you're in the UK and you ask Siri about a German psychologist – 'Hey Siri, tell me about Hans Eyesenck' – after performing its analysis, Siri may decide that it heard one of these:

> ***Hey Siri, tell me about ISEQ***
>
> ***Hey Siri, tell me about harms I think***
>
> ***Hey Siri, tell me about Hahnes I sing***

Not quite right. What's needed is yet more data, and better-trained machine-learning models. But there appear to be major downsides to this too. In 2019 a team of researchers analysed the environmental impact of training several of the most successful deep neural networks for natural language processing. They found that (in addition to a cloud compute cost of several million dollars) the carbon footprint could be as high as five automobiles for their entire lifetimes. While the AI technology may work efficiently once trained, the processes of creating the AIs are not efficient or cheap. 'In general, much of the latest research in AI neglects efficiency,' explains Carlos Gómez-Rodríguez, 'as very large neural networks have been found to be useful for a variety of tasks, and companies and institutions that have abundant access to computational resources can leverage this to obtain a competitive advantage.'

One option explored by Microsoft in 2016 was to use crowd sourcing to help provide data and let their Twitter chatbot learn. Tay was launched on 23 March 2016, but was hurriedly shut down just sixteen hours later when internet trolls taught Tay a variety of foul and drug-related language, which it then happily tweeted to its many followers.

There are yet other repercussions of the use of AIs in communication. Today, our technology can read through millions of social media posts and classify us into different categories. Millions of news articles and blogs

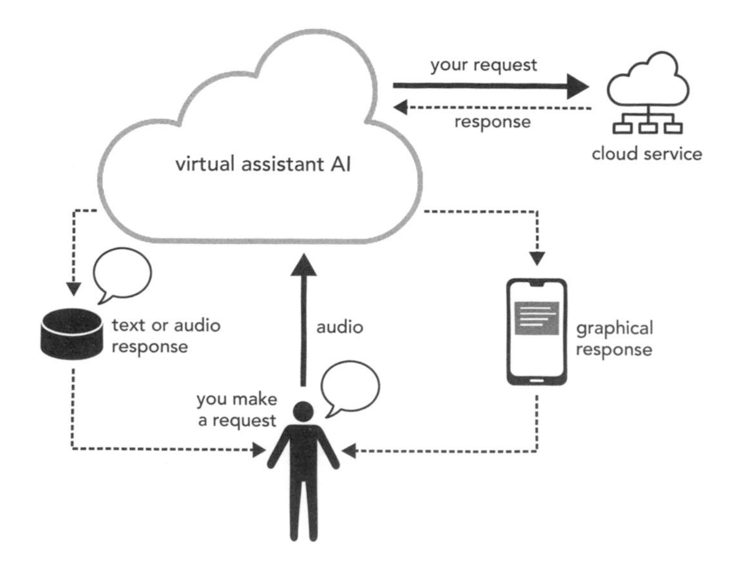

can be analysed daily to track public sentiment on specific topics. Targeted advertising misinformation or political messages can be produced by chatbots pretending to be humans. Opinions of populations can be monitored and managed. Even the way we are kept informed is curated by AIs. Recommender systems monitor the content of the news articles we like to read on our mobile devices, and feed us more of what we like, causing us to see a narrower view of the world and have our prejudices reinforced. Today it is easy for unpleasant regimes to control their populations, or for popularist leaders to gain power. Has AI helped to damage democracy and destabilize our world?

> **We've reached a fever pitch in all things AI. Now it's time to step back to see where it's going.**
>
> AMY WEBB
> NYU professor and futurist (2019)

Despite the bad headlines, AI technologies that work with our written and spoken languages have also transformed our world for the better, and still have the potential to do more good than harm. It is through natural language processing that researchers can now collate together thousands of separate scientific papers and derive new findings that no human could achieve. It may be through AI that we can understand the opinions and views of the millions of individuals in our populations and help enable our politicians and organizations to meet their needs better. All new technologies can be used for good or bad purposes. We need to recognize the impact of AI and ensure it is used appropriately.

08 RE-IMAGINE REALITY

'The true sign of intelligence is not knowledge but imagination.'

ALBERT EINSTEIN

Swirling, glittering clouds dance in blackness. They rush towards each other and seemingly tear each other apart into smoke, which quickly coalesces back into brighter sparks. They spiral around each other, fireworks with a mind of their own. A large spiral cloud filled with glitter forms, while other smaller spiral clouds drift past. But suddenly a new spiral cloud drifts too close, and like magnets attracting, the spirals catastrophically combine, scattering debris around them, their fiery internal cores briefly oscillating around each other until they combine into one. Now the spiral is rebuilt and the combined form continues its swirl in the blackness.

This is a view of the formation of a massive 'late-type' star-forming disc galaxy. Despite the vivid realism of the animation, no cameras could ever exist to capture such a sight as the duration of what it depicts spanned countless millions of years. But this is no pretty animation from the mind of a storyteller. This is the output from the IllustrisTNG project – a computer simulation that models the formation of our universe in more detail than has ever been achieved before, using our best understanding of the laws of physics.

Digital investigations

Artificial intelligence comes in many flavours. While many researchers attempt to engineer intelligence using their ideas of what might be the best and most effective way to solve a problem, others prefer to use computers as scientific tools in order to carry out investigations. Building a computer simulation is not the same as making a computer animation or game, where anything goes. The aim in simulation is to make a virtual laboratory, which behaves exactly according to reality – except that we are in control. If we wish to understand something too slow to see in real life, we can speed up time in our simulation. If we wish to delve inside a complex form that might be impossible to dissect in real life, we can use the computer to tell us the precise internal constituents. If we wish to ask 'what if' questions, then we can subtly modify the simulation: what if the laws of physics worked a little differently? What if evolving life faced a much hotter environment?

Modelling reality is not easy, and relies on very careful collection and usage of data. There's an old saying, 'Garbage in, garbage out' (often abbreviated to GIGO). If your model is incorrect then you cannot expect its predictions to be accurate. All models are wrong by design, as our computers are not powerful enough to simulate every aspect of reality. The trick is to model those aspects of reality that produce the behaviour of interest, and omit everything else that has little effect. Researchers must therefore attempt to create abstractions and simplifications of reality that are powerful enough to provide us with new answers, but simple enough that computers can actually run the simulations in a practical amount of time. They must carefully calibrate each part of their models, ensuring that it behaves correctly according to real-world data. Any predictions made by the model that are incorrect should also be used to refine the model further. A common criticism of economic models is that they are rarely validated with real data, and often make assumptions that people will behave rationally. This can lead to wildly inaccurate predictions that are of little help when attempting to stabilize or regulate volatile and chaotic markets.

' Garbage in, garbage out. Or rather more felicitously: the tree of nonsense is watered with error, and from its branches swing the pumpkins of disaster. '

NICK HARKAWAY
The Gone-Away World (2008)

Many of the AI methods in previous chapters can be considered as types of computer simulation. Karl Sims' virtual creatures were a simple simulation of evolving life in a simulated environment. Early ideas of neural networks were based on understandings of how neurons worked. Many optimization algorithms are based on simple ideas of how living systems work: genetic algorithms are inspired by natural evolution, ant-colony optimization is based on the way ants collectively find shortest paths from nest to food, artificial immune system algorithms are based on the way our immune cells detect and respond to pathogens. Machine-learning algorithms create simple 'models' of the data they learn, which can be used for prediction.

There are also other techniques used specifically for modelling and simulation. Models may be based on many equations used together in combination; perhaps a set of differential equations that are solved in order to determine the behaviour of an electrical circuit or chemical processes. Spatial models (which try to determine how physical objects may behave in the real world) may typically divide space into little pieces known as a mesh and calculate the state within each element of the mesh, perhaps by following equations. Computational Fluid Dynamics uses this approach to model the complex behaviours of gases and liquids, enabling aircraft designs to be refined in simulation, and methods such as Finite Element Analysis are commonly used to understand stresses on structures before they are built, allowing us to design safer structures using the right materials.

Cellular automata

A simple cousin of FEM known as cellular automata has been used for AI for many decades to simulate everything from chemical interactions to physics simulations. A basic CA is a simple grid of cells that may be filled or not filled. When the simulation runs, time is chopped into discrete steps, and for each timestep the computer iterates through every cell and follows simple rules, e.g., if two neighbouring cells are filled then this cell is filled, otherwise it is empty.

The concepts behind cellular automata were first proposed by mathematician Stanisław Ulam, who wanted to simulate an idea by John von Neumann to create a liquid containing electro-magnetic components. Both Ulam and von Neumann were fascinated with these ideas, leading to the latter proposing that living systems could be defined as things that can reproduce themselves and simulate a Turing

> **If people do not believe that mathematics is simple, it is only because they do not realize how complicated life is.**
>
> JOHN VON NEUMANN (1947)

CONWAY'S GAME OF LIFE

The Game of Life considers a filled cell in the grid as 'live' and an empty cell as 'dead'. It has four simple rules for its cellular automata:

Underpopulation: a live cell with fewer than two live neighbours dies.

Live: a live cell with two or three live neighbours stays alive.

Overpopulation: a live cell with more than three live neighbours dies.

Reproduction: an empty (dead) cell with exactly three live neighbours becomes a live cell.

Seed the CA with a few randomly placed live cells, run the simulation, and suddenly you have an organic-looking spread of growth, following weird patterns. Sometimes it may all die out, sometimes it may turn into little oscillating shapes that persist. Seed the CA with exactly the right complicated shape and it can even make copies of itself – you can have a von Neumann Universal Constructor within the Game of Life.

machine (perform all tasks accomplishable by computers). Von Neumann defined a 'universal constructor' as a machine that can make identical copies of itself by processing its environment, and he used a cellular automata algorithm to explain how it could work. In 1970, British mathematician John Conway took this idea and created a special kind of cellular automata, which became known as the Game of Life (see box opposite).

SEED SHAPE

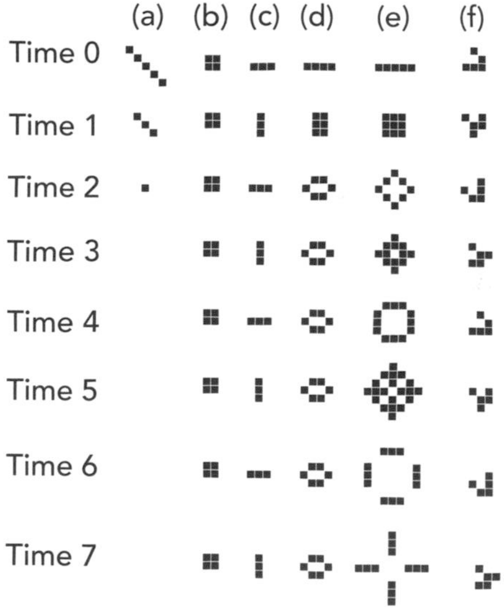

(a) (b) (c) (d) (e) (f)

Time 0

Time 1

Time 2

Time 3

Time 4

Time 5

Time 6

Time 7

Cellular automata were so loved by some scientists that they claimed that CAs could explain the workings of biological systems, and even our universe. Stephen Wolfram, computer scientist and creator of the popular Mathematica software, is one such person. 'It is perhaps a little humbling to discover that we as humans are in effect computationally no more capable than cellular automata with very simple rules,' he says. 'But the Principle of Computational Equivalence also implies that the same is ultimately true of our whole universe.'

Digital agents

While cellular automata discretize space and time into little chunks, other kinds of model are more relaxed. Agent-based modelling (ABM) emerged from ideas such as von Neumann's work on cellular automata and eventually developed into a scientific methodology in its own right. (The similar-sounding multi-agent computing is another related AI approach that looks at how software agents interact to solve problems, but is less concerned about modelling and focuses more on practical problem solving.)

Agent-based models (ABMs), sometimes known as individual-based models, are a class of algorithm that simulate the behaviour of autonomous entities, known as agents. These might be biological cells in an organ, molecules in a liquid, people in a population, or any collection of similar entities that have independent behaviours that depend on interactions with each other. ABMs are used to study a larger system, creating a virtual lab within which they are released to do their thing. An ABM uses rules or algorithms for each agent, enabling it to act autonomously and interact with its companions. ABMs may use any form of AI to drive agent behaviours, including some of the more complex deep learning algorithms. Even if the behaviours of each agent are driven by relatively simple rules or algorithms, when they interact and affect each other with a bit of randomness thrown in for realism, the result is an emergent and frequently unexpected higher-level behaviour. One flying bird appears quite simple, but a huge flock of birds can swirl and instantly change form as though the whole has some greater intelligence. ABMs can bring together many aspects of AI and related fields such as complex systems, evolutionary computation, economics, game theory, sociology and even psychology.

Craig Reynolds is a computer graphics expert and one of the scene programmers for the original *Tron* movie. In 1986 he created an ABM algorithm, which he called boids (shortened from bird-oid objects, meaning bird-like objects). His algorithm was the first to illustrate how independently moving agents, each following simple behavioural rules, could produce exactly the same bird-flocking and fish-schooling behaviours observed in the natural world.

The base rules were remarkably simple: each boid tried to avoid colliding with neighbours, flew in the average

direction of neighbouring boids, and moved towards the centre of mass (average position) of the boids nearby. Reynolds' algorithm worked so well that it has been used in the movie industry ever since to simulate flocks of birds or crowds, one of the first examples being computer-generated bat and penguin colonies in the film *Batman Returns* (1992). The same kinds of algorithms are now commonly used to control swarms of robots to ensure they work together effectively as they solve a group task. Reynolds continues his work on simulating agents to this day by contributing to Righthook – a company developing autonomous robots and vehicles.

Shortly after Reynolds had created boids, a new kind of AI was formally named: artificial life. This field combined the interests of AI researchers with

> **Flocking is a particularly evocative example of emergence: where complex global behaviour can arise from the interaction of simple local rules.**
>
> CRAIG REYNOLDS (2001)

CHRISTOPHER LANGTON (1949–)

Computer scientist Christopher Langton invented the term artificial life, and created the first ever workshop on the Synthesis and Simulation of Living Systems in 1987, which subsequently became known as the Artificial Life conference. Chris pioneered early studies of complexity arising in cellular automata, and suggested that complex forms, especially living systems, existed in a region between order (where everything is static and regular) and disorder (where everything is random). It was an important new way of looking at living systems: instead of regarding them as akin to clockwork machines that operated with predictable certainty, life lies 'at the edge of chaos', its components interacting such that they become more than the sum of their parts. Emergent behaviours arise that are unpredictable, yet desirable. Langton created some deceptively simple models such as the 'Langton Ant', which moved according to very simple rules, leaving a trail behind it, and the 'Langton Loop', which modelled a very simple kind of artificial life that had its own heritable genetic information. Despite their seeming simplicity, the resulting emergent behaviours gave significant insights into the development of complex forms.

biologists, chemists, philosophers and even artists – all those who wanted to use computers to investigate fundamental questions about living systems. ALife researchers typically use ABMs to explore how the first self-replicating molecules arose, how cells developed, how multicellular organisms evolved, how brains and perception formed, and how complex ecosystems work. While other forms of AI such as deep learning focus on engineering more effective solutions, in ALife and related fields of computational biology, researchers investigate models that match biology more closely, to understand how they work (for example, spiking neural networks; see chapter 10). Computer simulation is essential for such extraordinary research.

Virtual futures

Computer simulation and modelling is the imagination of AI, and imagination is one of the most powerful forms of intelligence that we possess. We go beyond simply predicting that the toast we just dropped will hit the floor. We can imagine entire scenarios, entire worlds, with imaginary characters having imaginary behaviours. Computers have far better imaginations. With the right algorithms, they can imagine entire universes.

Today these models are used to simulate the motion of crowds so that we can design better buildings, they're used to predict how tumour cells might react to cancer therapies, how economies may change over time; they're even used to investigate the origin of humans and how our societies formed. Simulations are used extensively in the entertainment industry for movie special effects, virtual and augmented realities, and computer games. They are used every day to predict the weather, and there have been notable successes where simulations have been used to prevent disease. For example, in 2001 there was an epidemic of foot and mouth disease in the UK. A simulation predicted that a significant cull of livestock would transform an exponential growth of the disease to exponential decay within two days. The recommendation from the model was implemented, and although it meant a tragic loss of life, it worked, and the disease was eradicated.

But computer simulation is typically one of the most computationally intensive forms of AI we undertake today. Even if its input data is correct, we are limited by the computational resources available. Despite accelerated hardware, our computers cannot simulate reality in the detail we'd like, and simulations may take infeasible time to run.

> **"The art is to find an approximation simple enough to be computable, but not so simple that you lose the useful detail."**
>
> MICHAEL LEVITT
> Nobel-prizewinning Professor of Structural Biology (2013)

Some simulations are so important that they may affect the future of life on Earth. Climate models are among the most complex of all simulations. They must integrate data from vast numbers of historical sources and current sensors and make predictions about how the concentration of various gases in our atmosphere may affect global warming and weather patterns in the future. As we perfect them, we better understand how to prevent damaging our planet further. While it is always tempting for researchers to model everything possible, and use the all available computational resources, sometimes the complexity can be so great that results are hard to interpret. It is therefore essential to have models (for example those that use probabilistic methods) that do not just make predictions, they provide a level of certainty for those predictions, so that we can understand how much to trust them.

09 FEEL BETTER

'But feelings can't be ignored, no matter how unjust or ungrateful they seem.'

ANNE FRANK

Elliot had a successful job in a business firm; he had a nice life with friends and family. He was a role model to his siblings and a good husband. But Elliot started having severe headaches. After tests, he was diagnosed with a brain tumour in his prefrontal lobe. Elliot was lucky, for his tumour was benign and operable, and it was successfully removed. He was well again, and any minor damage to his brain from the surgery seemed to have no effect. He was able to go back to his job, and continue his life as though nothing had changed. But as the months went by, Elliot's life slowly fell apart. Now, when he tried to take on new projects, he never finished them. He made countless mistakes that had to be corrected by others. He was fired from his job, and before long his marriage also broke up. He got involved with a dubious new business partner on a bad money-making scheme and went bankrupt.

He married again, to a woman who seemed entirely unsuitable for him; the second marriage quickly broke up. He eventually moved in with a sibling, who realized something was wrong. Elliot was no longer quite the same person he used to be.

He was referred to neuroscientist Dr Antonio Damasio. On examination, Elliot appeared perfectly normal. He had excellent memory, no speech impairment, nothing visibly wrong. But he could no longer *feel*. Elliot had no emotions. He was the perfect, logical decision-maker – except that, without emotions, he couldn't decide anything. Faced with countless decisions every day, he had no way to assign a positive or negative judgement. Should he eat cereal or toast? Should he categorize his documents at work by date or by importance? What should he do this evening? Or next week? Elliot simply drowned in the overwhelming sea of decisions and ended up either frozen in indecision, or choosing badly, ruining his life.

Elliot was not the only known example. In the 1930s a neurologist had discovered a stockbroker who also

‘ I began to think that the cold-bloodedness of Elliot's reasoning prevented him from assigning different values to different options and made his decision-making landscape hopelessly flat. ’

ANTONIO DAMASIO (2005)

sustained damage to his prefrontal lobe. Once again, the man's life was ruined. He never left his house, just stayed home plotting his professional comeback and boasting of his sexual prowess, despite no longer working or having sexual partners. In the same decade, a poorly thought-out surgery on a different patient with anxiety and schizophrenia had similar results. There could be no doubt – emotions help us to make decisions. Without them, while we might have a high IQ, and an ability to solve problems, we cannot function in the real world.

Emotional intelligence

These ideas are relatively new and of growing importance to artificial intelligence. Throughout history, most AI researchers have been male scientists and engineers – not necessarily known for their empathy or social graces. During decades of research, almost all ambitions were to achieve cold, logical decision making, control and prediction in AI. The idea that an AI should understand emotions, or even have emotions, was not common – indeed most researchers believed that emotions got in the way of decision making, and it was better to have none.

Some of the earliest work in Emotion AI, or affective computing as it is often known, began in 1995 by Rosalind Picard, a professor at MIT's Media Lab. Picard was directly influenced by Damasio's work with Elliot, and the revelation that emotions are essential for decision making. The researchers in her lab have since spent two decades exploring all aspects of emotional AI.

Some of the initial successes came with the advent of improved facial and speech recognition (see chapters 5 and 7). Picard and her researchers discovered that in addition to recognizing faces, the AIs could be trained to recognize emotions from the expressions on faces. In addition to recognizing speech, the AIs could be taught to detect changes in tone that correspond to nerves, or anger, or even dishonesty. Picard founded Affectiva with one of her PhD students, Rana el Kaliouby, in 2009. The company pioneered machine-learning methods to analyse the emotions of people as they watched advertisements, enabling more than 25 per cent of the Fortune 500 companies to understand exactly what effect their adverts have on people, so they can be carefully improved. They also created solutions for vehicles that monitor the emotional state of occupants using facial and voice recognition in order to keep people safer on the roads and ensure their comfort in the vehicles. Today, facial recognition methods are also used for the identification of autism,

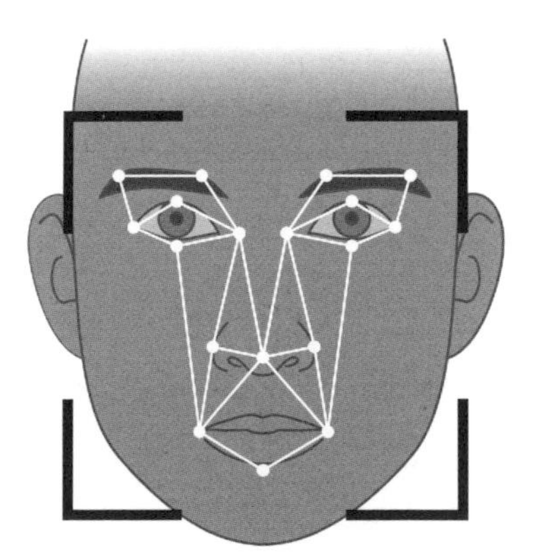

schizophrenia, Alzheimer's disease and within crime prediction systems.

FACIAL EMOTION RECOGNITION

Facial emotions are recognized by using machine learning to identify the results of individual muscle movements, or 'action units' (AUs). A Facial Action Coding System (FACS) is then used by the computer to decipher what emotion is being displayed. For example, AU6 + AU12 + AU25 suggests 'happy' while AU4 + AU15 + AU17 corresponds to 'sad'. Some AUs frequently happen at the same time, some only happen occasionally and might change the meaning of the emotion (think of a frown with a smile). Some rarely happen together at all, such as AU25 (lips apart) and AU24 (pressed lips). Using these rules, derived from many years of research, the computer can analyse images and video of faces and detect even transitory micro-expressions that we might find hard to spot. It can also tell the difference between a fake 'PanAm' smile and a sincere and 'Duchenne' smile.

Another MIT spin-off called Empatica uses wearable wrist sensors to help monitor individuals susceptible to seizures. The researchers developed machine-learning approaches to classify neurological events as being due to emotional stress (which can then inform us how to reduce stress in our lives) or to serious events such as seizures (which could help bring appropriate medical care quickly). 'While measuring the physiology of emotion we learned that some neurological events – several kinds of seizures – happen in regions deep within the brain that are involved in emotion,' explains Rosalind Picard. 'These events can be measured from the wristbands we built that were originally designed to measure aspects of emotion.'

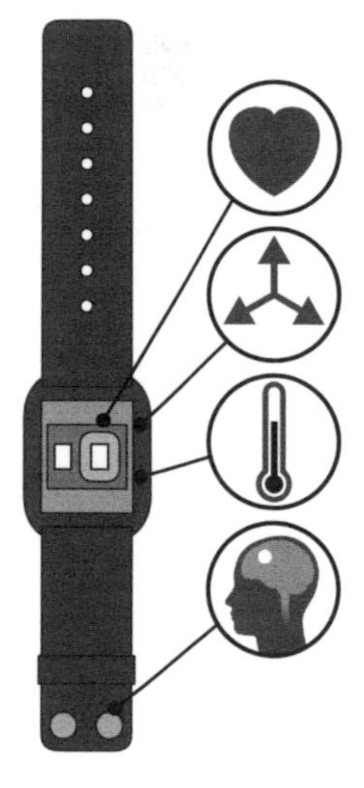

Such technology is now being taken up in many businesses. For example, the Amsterdam-based companies Koninklijke Philips Electronics and ABN AMRO Bank placed wearable sensors on the arms of financial traders. Machine learning detected and classified emotional states, relaying them in the form of colours to the wearer. When traders become more aware of their own emotions, the trading risk is reduced, preventing accidental – and extremely costly – errors.

Social robots

AIs may be able to recognize emotions, but building intelligent robots that interact with us has always caused problems for roboticists. We can become remarkably uncomfortable when an AI attempts to control lifelike human- or animal-like forms. Weirdly, we're happy with robots that resemble toys or cartoon characters, and of course we're happy with reality, yet when we see a semi-realistic robot that's not quite right (as is the case with all current robots) then we find them almost like animated corpses or zombies. This effect is known as the uncanny valley effect. As realism increases to a point where the device is nearly lifelike, we feel uneasy. As soon as the robot moves, our emotional response becomes even more negative.

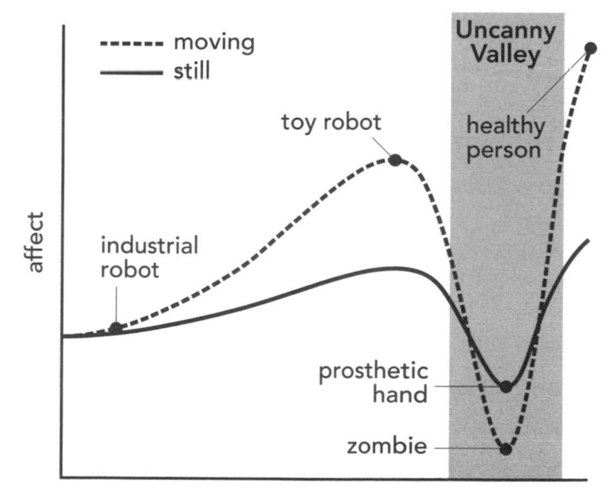

This is a problem, because there are countless applications for which robots need to interact with people. It would be extremely useful to have robots that could monitor the health of people in hospitals, or the elderly at home. It would be great to have robots that took over boring receptionist jobs or provided information to shoppers in shopping malls. These robots ideally should provide friendly companionship to us, make us happy, and be aware of the emotional impact they have on us. Such robots are known as social robots.

One of the earliest social robots was Paro the seal. Designed by Takanori Shibata at the Japanese National Institute of Advanced Industrial Science and Technology in 1993, Paro was modelled after real harp seals that Shibata observed in north-eastern Canada. The little snow-white fluffy robot subtly moves its limbs and body, and makes seal noises using the recording of the seals taken by Shibata. Paro was designed to respond to being stroked, to look into our eyes, respond to its name and learn actions that generate a favourable reaction. Shibata avoided the uncanny valley effect by making Paro very toy-like. It eventually was registered as a medical device and was used for therapy in care homes for patients with dementia.

Universities have led many advances in the area of social robots, with MIT's Cynthia Breazeal pioneering Kismet – a slightly freaky-looking cartoon-like robot head able to understand facial expressions, gestures and tone of voice, and express its responses with moveable lips, eyebrows, eyes and ears. Kerstin Dautenhahn and Ben Robins of the University of Hertfordshire developed Kaspar, a child-like humanoid robot that is used to interact with autistic children and help them with communication, interpersonal skills, play, emotional wellbeing and development of preschool skills. 'Many autistic people are drawn to technology, particularly the predictability it provides, which means it can be a very useful means of engaging children, and adults too,' explains Carol Povey, director of the National Autistic Society's Centre for Autism.

Companies have for some time attempted to create home robots that engage users with more emotional awareness and personality, such as the baby dinosaur robot Pleo, the Vector Robot by Anki, or the Keepon – a cute little dancing yellow snowman-shaped robot invented by Hideki Kozima while at the National Institute of Information and Communications Technology (a version of which, called Zingy, appeared in an advertising campaign for EDF Energy in 2012). Back in 2006,

Nabaztag was a smart home robot in the shape of a sculptural rabbit, which could read emails, express some amusing random thoughts – 'Where is my hat? Have you seen my hat?' – and move its ears. Despite a research study showing it was popular for companionship among elderly users, it never worked well enough to satisfy customers and the company did not survive – a trend seen by many other social robot companies. It is perhaps sad that these robots, which are arguably the nicest, most fun, and of most benefit to users, were by 2020 mostly replaced by endless home hubs. Sold at a loss because they were used as marketing tools to increase purchasing in the home, these anonymous, cylinder-shaped speakers might sound as though they could express emotions through the tone of their voices, but they had zero understanding of our emotions, and no real emotions themselves.

Emotional robots

The ultimate aim of Emotion AI or affective computing is to give AI its own emotions. Recognizing emotions in others only allows you to understand that your behaviour is having some effect. Feeling your own emotions enables you to empathize, and make decisions more effectively.

Researchers have simulated emotions in their AIs and multi-agent models for many years. Some simply use the concept of a value (for example, happiness is 0.9); some use a fuzzy logic model (see box on page 144). Simulated emotions may be reactive – fear triggered by a fire, happiness by food. Some models of emotion incorporate psychological features such as personalities to moderate how each emotion effects behaviour. Symbolic AI methods have been used to create rules about emotional states: 'If the appraised situation has raised clarity and pleasure, it creates joy', enabling the agent to 'feel' joy, anger, surprise, fear, panic, anxiety and regret, and reason about these emotions. Other researchers are now starting to simulate physiological implementations of emotions, with emotional states modifying the behaviour of neural networks just as brain regions known as the limbic system and amygdala may affect the overall behaviour of our own brains. In the same way that real emotions help us to function in the real world, so the researchers hope that incorporating emotions into artificial neural networks will enable our robots to mirror and learn emotional states, giving them their own ability to make quick and effective judgements, like us.

> **The question is not whether intelligent machines can have any emotions, but whether machines can be intelligent without any emotions.**
>
> MARVIN MINKSY (1986)

FUZZY LOGIC

Fuzzy logic is a kind of multi-valued logic first proposed by pioneering mathematician and computer scientist Lotfi Zadeh in 1965. While normal logic is all about binary yes/no, on/off values, fuzzy logic defines a degree of membership in a set. Fuzzy logic resembles, but is distinct from, probability: a fuzzy logic value represents the degree to which an observation is within a vaguely defined set, while probability represents the likelihood of something. While a logical expression might have binary logic variable *happy*, which might take the value *yes* or *no*, in fuzzy logic the same concept would be formulated using several linguistic variables corresponding to overlapping fuzzy sets, for example, the concept *mood* might be fuzzified into a value of 0.8 for membership of the fuzzy set *happy*, 0.3 for the fuzzy set *sad*, 0.1 for *anger* and 0.4 for *surprise*. The use of linguistic variables makes fuzzy logic much more understandable, and its ability to represent logic using multiple values means that it can be used for control applications, providing much greater precision. Today, fuzzy logic is widely used in the AIs that control underground trains, lifts and even rice cookers.

Caring AI

Affective computing is one of the latest big advances in AI, with the affective computing market growing from an estimated USD 22.2 billion in 2019 to USD 90 billion by 2024. The increased use of wearable technologies makes gathering data (images of faces, audio of voices, and physiological data such as heart rates and perspiration) more and more easy. Our machine-learning algorithms are now more than capable of classifying these data and enabling our emotional state to be accurately predicted. This should lead to more aware chatbots that moderate their responses to us when we are clearly angry or frustrated. Car manufacturers are already working on vehicles that warn us about driving when our faculties are adversely affected by anger or depression. Better sentiment analysis is already in use by the British government to map social media comments to specific emotional sentiments. As digital visors (such as Google Glass) become more common, companies like Brain Power are introducing new aids for autistic people, giving them cues on the emotional state of others around them and teaching them to understand the emotional world better. Other companies are producing solutions to help with mental health, using behavioural therapy delivered via chatbots –

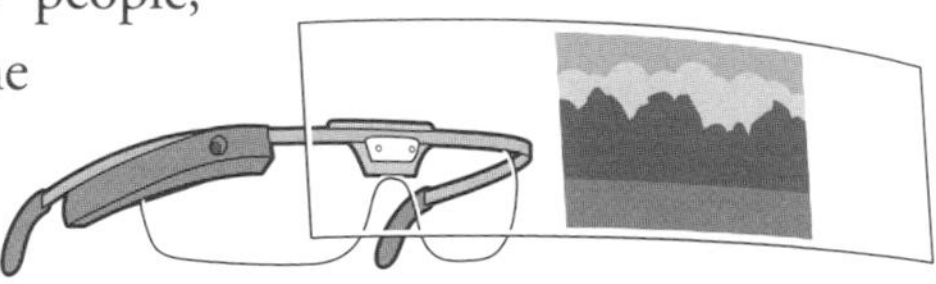

'Ellie' is an American DARPA-funded virtual agent that helps soldiers with PTSD. 'Karim' is a chatbot developed by Silicon Valley start-up X2AI that helps Syrian refugees overcome trauma.

The future looks bright. However, as with all advances, emotional AIs can have their dark side. The use of our physiological and emotional data without our permission is causing many to have concerns over privacy. And what if the data is used for purposes that are less pleasant? Microsoft's female chatbot Xiaoice achieved a sizable following because of her pretty looks and ability to charm those who chat with her. With around 100 million eighteen- to twenty-five-year-olds chatting to her, some on a regular basis as a friend-substitute, is this the kind of cure for loneliness we should be using? Or is this a cynical ploy by tech companies to engender brand loyalty?

> **By 2022, your personal device will know more about your emotional state than your own family.**
>
> ANNETTE ZIMMERMANN
> research Vice President at Gartner

In 2017, leaked Facebook memos sent to advertisers claimed that the company could monitor posts in real time and identify when teenagers feel 'insecure', 'worthless' and 'need a confidence boost'. They stated that they could classify feelings such as 'stressed', 'defeated', 'overwhelmed', 'anxious', 'nervous', 'stupid', 'silly', 'useless' and a 'failure'. Facebook was also criticized for earlier research in 2014 during which they manipulated the news feeds of nearly 700,000 users to affect their emotions, breaching ethical guidelines, as none had given informed consent.

We would never allow a physical robot to run around flailing randomly at people, as it would inevitably cause injury. It is clear that there is still a lack of recognition that Emotional AIs could also cause damage – and emotional damage is much harder to detect and repair. As we incorporate emotions into the awareness and capabilities of AI, we must not treat Emotion AI as a new tool with which to manipulate people. This could cause the most harm of all, both to individuals and societies.

10 KNOW YOURSELF

'To say that AI will start doing what it wants for its own purposes is like saying a calculator will start making its own calculations.'

OREN ETZIONI

At the time of his death in February 1988, world-renowned physicist Richard Feynman left a message behind. The words were chalked in his somewhat uneven handwriting, on the top left of his Caltech blackboard. These words have inspired many computer scientists for decades, and continue to do so today. Eight simple words, which have been used to justify entire research programmes worth millions.

'What I cannot create, I do not understand.'

Owen Holland is a scientist who identifies with these words. Throughout his career, Holland has explored how biological systems work, attempting to unpick the mysteries of intelligence through building his own equivalents, pioneering many fascinating robots along the way. (One of his more unusual robots was the Slugbot – a robot that 'ate' real slugs and aimed to generate its power from their rotting bodies.) His work culminated in an ambition to understand consciousness – perhaps the ultimate dream in artificial intelligence. Holland's approach was to mimic nature more than had ever been attempted before in robotics. He and his team built CRONOS, a rather scary-looking humanoid robot with a single cyclops eye in its head, but with anatomy matching a human skeleton as much as possible. Bones were carefully shaped in plastic to match human bones. Actuators combined with stretchy tendons pulled on the bones, behaving like muscles. Holland believed there was a link between the complexity of a body and the complexity of a brain needed to control it.

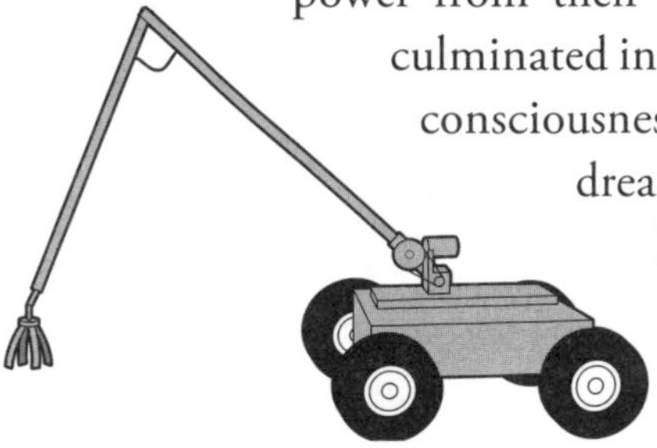

The body of CRONOS was extremely complex. In fact, CRONOS was so complex and difficult to control that ultimately Holland and the researchers that worked with him discovered they couldn't do it. CRONOS might have the body of a conscious robot, but the researchers struggled to make a conscious brain to match and spent most of the time trying to engineer working control systems. (The lessons learned from this work resulted in new actuators and new robots such as Roboy – a slightly more friendly child-size version.)

But all was not lost. During the course of the research, Holland believed he found some of the answers. He believed that consciousness was all about the robot forming its own internal model of the outside world – its own imagined reality – which should include a model of itself. When a robot could think about itself, its own body and how it might react and affect the real world separately from the world itself, then perhaps this self-awareness would be the first step towards consciousness. Step by step, new features could be added – language, speech, memory, motivations – while the researchers looked inside the mind of the robot and saw what it was thinking. Perhaps when the robot's own imagined idea of itself started to become more important than reality, it would become conscious.

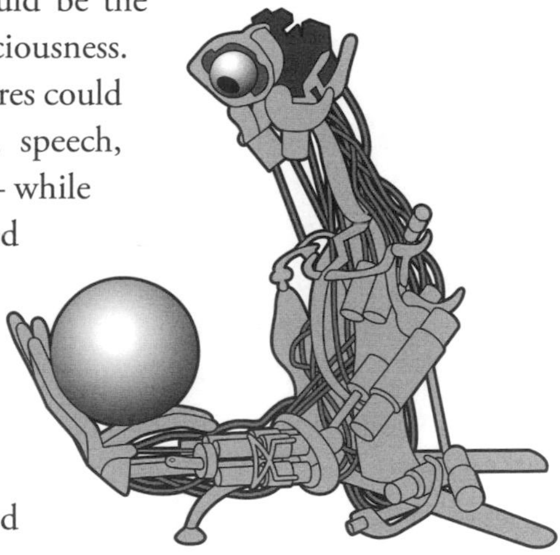

Holland remains optimistic that we will achieve computer consciousness in the coming decades. 'You remember the stickers on computers that said Intel Inside,' says Holland. 'We're waiting for the stickers that say Consciousness Inside.'

Artificial general intelligence

While researchers such as Holland aim for better self-awareness by computers – an ambition that could improve AI's ability to understand itself and its actions better – a conscious AI does not necessarily imply a useful AI. Perhaps not even a clever AI. For a more widely useful kind of AI, some believe we need a more general kind of intelligence.

Today we do not have many AI generalists. Almost without exception, today's AIs are specialists, experts in very narrow domains. Ask a quality-control AI to tell you about the weather tomorrow, or ask the music-generating AI to predict share prices, and you will not get any coherent answer. It's more efficient and practical to make AIs with narrow expertise, which is why the majority of AI solutions take this approach. But when we want our AI to interact with us in more general ways, to understand our needs, have real conversations, or to understand wider social, ethical or environmental repercussions of their behaviour, we start to think about a more general kind of AI.

Artificial general intelligence is the term that some researchers use to describe this stronger form of AI. It encompasses the grand vision of creating AI systems possessing general intelligence at the human level and beyond. A large diversity of techniques is used in combination in an attempt to enable the AI to gain a greater set of capabilities. Some aim to combine the symbolic or model-based approaches with sub-symbolic methods such as deep learning. Some try to create giant knowledge graphs to enable broad general knowledge.

Despite decades of researchers trying and many companies claiming to be aiming towards this goal, it is clear that progress is slow. One method for measuring the level of human intelligence is simply to give the AIs an IQ test. In 2017, the IQ of freely available AIs such as Google AI or Apple's Siri and others was 47 at best, equivalent to a six-year-old child. But other researchers argue that the same AIs would not be able to walk, converse, or even make a simple cup of coffee, meaning that the goal of general AI was not being met.

Simulating brains

While engineering AGI may be the choice for many computer scientists, several major efforts in modelling and simulation of biological brains provide a different route. The Human Brain Project, started in 2011, is a hugely ambitious project that will cost an estimated 1.019 billion euros during its ten-year lifetime. The aim is to investigate and simulate the human brain at all levels, from the microscopic to the entire brain. It's a controversial aim that most researchers feel is not feasible with today's technology or knowledge, but those working in the project hope that, at the very least, benefits to neuroscience and AI will be achieved as they pursue such a grand vision.

This vast project is just one of many. The American BRAIN Initiative also began in 2013, funded by the Obama administration, with aims to progress neuroscience and understanding of brain disorders through advanced scanning and modelling. The similarly motivated China Brain Project was launched in 2015, and should run for fifteen years. Computer scientists are even building massive neural computers, designed specifically to simulate spiking neural networks.

Compared to artificial neural networks as used in machine learning, spiking neural networks resemble biological neurons a little bit more. Traditional artificial neurons output a continuous value based on a function of their inputs. Spiking neurons, when activated by spikes on their inputs, fire chains of spikes at each other, effectively encoding information over time in binary on/off signals. This means that spiking neural networks may be better suited for handling problems that change over time, compared to conventional artificial neural networks. SNNs require heavy computation to simulate more realistic neurons (hence the need for

SPINNAKER

SpiNNaker is the brainchild of British computer pioneer Steve Furber, who created a new computer architecture inspired by the human brain, for use in neuroscience, robotics and computer science.

SpiNNaker comprises a dedicated set of processors that simulate spiking neurons. Each processor comprises eighteen smaller processors (sixteen used to simulate neurons, one for management and one spare) with a clever design to ensure they can communicate at ultra-high speed with each other and with their companions on neighbouring chips. The architecture allows many such chips to be used together in parallel in order to simulate the parallel working of billions of neurons, with an eventual target of 1,036,800 processors, which would require 100kW power – surprisingly small for such a mammoth parallel computer. SpiNNaker is one of the major hardware platforms forming part of the Human Brain Project. In addition to its use in simulating spiking neural networks, the researchers hope that their work will aid in the development of new types of energy-efficient massively parallel computers in the future.

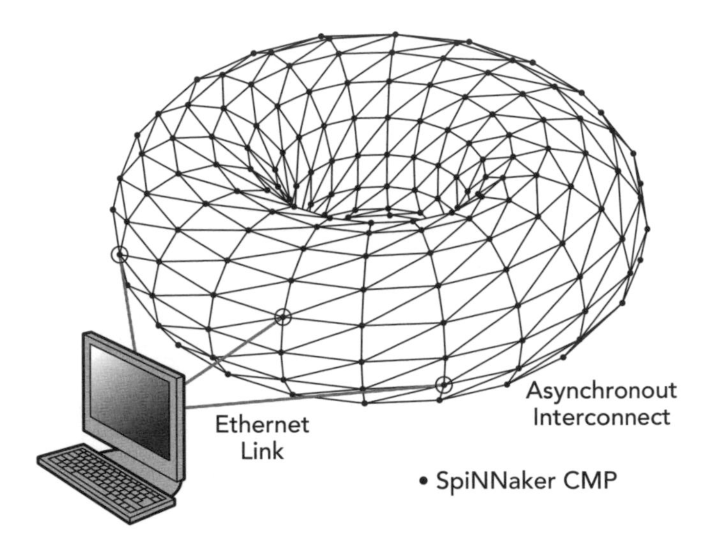

dedicated hardware such as SpiNNaker and others). It is also unclear how to make them learn, for traditional backpropagation cannot be used. For this reason, although spiking neural networks have great potential – including for neuroscience models to understand real brains better – currently we do not know fully how to make them work.

Other researchers continue to innovate in the area. Julian Miller pioneered the evolutionary approach known as Cartesian Genetic Programming to enable computers to evolve electronic circuits. Today he is inspired by the vision of creating artificial general intelligence by using

CGP to evolve novel neural networks. Unlike most neural networks, which are trained to solve a single task and must be retrained before they can solve others, Miller creates one artificial brain that can solve multiple tasks from a small number of examples. His evolved neural networks vary the number of neurons during learning and reuse neurons for different problems.

> **I am working on the moonshot challenge of AGI. I think we need to investigate much more complex models of neurons. In particular, neurons that can replicate and die.**
>
> JULIAN MILLER (2019)

To infinity and beyond!

Not everyone is optimistic about the benefits of artificial general intelligence. Some futurists, philosophers and science-fiction writers make some frightening predictions about where such technology might ultimately lead. They claim that as soon as an AI is clever enough to start designing itself, you have a runaway feedback loop – suddenly the clever AI could make an even cleverer one, which could make an even cleverer AI, and so on. This argument is often justified with reference to Moore's Law, which states that the number of transistors in a chip roughly doubles every two years (and leads to the corollary that the speed of processors will double every eighteen months). With computers becoming exponentially faster, super-intelligent

AIs will surely become an inevitability, say proponents of the idea. We will someday reach a 'technological singularity' where the growth of such AIs becomes uncontrollable, leading to potential disaster for the human race.

"I set the date for the singularity – representing a profound and disruptive transformation in human capability – as 2045. The non-biological intelligence created in that year will be one billion times more powerful than all human intelligence today."

RAY KURZWEIL
inventor and futurist (2005)

Such ideas are very entertaining in movies and science-fiction books, but should be taken with a pinch of salt in the real world. The development of AI technology has benefited from the growth of computation power, and the new availability of vast datasets to train machine-learning algorithms. But advances in AI are always because of algorithm developments, not processor speeds. (And AI winters are caused by disappointments when the algorithms cannot match the hype.) Each algorithm provides exciting new possibilities, but no single algorithm will solve all problems. New problems require new algorithms to solve them (just as the various tasks we undertake make use of different regions of our brain), and the pace of innovation is limited by our ability to create new algorithms, as well as our ability to understand the nature of intelligence itself.

It would be great if we could use a shortcut and let computers design themselves. But despite methods such as genetic programming, we struggle to make computers design better algorithms on their own (and why would they, without being directed to do so?). Today we have thousands of human scientists making use of the very latest computers as they attempt to make AI. If we are struggling to create intelligence, why would our considerably less clever AIs be any better?

> **This prior need to understand the basic science of cognition is where the "singularity is near" arguments fail to persuade us.**
>
> PAUL ALLEN
> co-founder of Microsoft (2011)

We do not even have the benefit of an exponential growth to help us. It turns out that although there may be an exponential, it's almost certainly not in our favour. Every time we attempt to make an AI just a tiny bit cleverer, we exponentially increase its capabilities. This produces exponentially more chances for it to fail, so we must run vastly more tests and experiments to ensure the design is right. Ultimately, evidence so far from AI research indicates that more complex capabilities mean slower, more difficult progress. We know from physics that it takes more and more energy to make progress as we approach the speed of light. Something similar seems to occur as intelligence increases – it becomes ever more difficult to increase the capability of intelligence. It may be no coincidence that there are very few creatures on planet Earth with brains as complex as ours, or that it took 3.5 billion years of evolutionary testing to create the few designs that exist.

We can be reassured that there will be no singularities, no runaway AIs. Our dedicated and hardworking computer scientists and engineers will continue their research, slowly inventing new algorithms to provide helpful AI tools to make our lives safer and easier. It's possible we will never achieve AI as advanced as the human brain. If we do, it may take geological timescales.

AI futures

Most AI and robot research and products have nothing to do with superintelligence or artificial general intelligence. The majority of AI is dedicated to finding focused solutions to specific problems: a new chatbot to take over mundane queries in call centres; new automated driving algorithms to help with motorway driving, or parking our cars; more advanced robots that automate more of the boring manufacturing processes; fault detection systems to ensure our factories produce error-free products; wearable computers that detect medical conditions earlier than ever before; social robots that provide comfort and care for the elderly or disadvantaged; internet bots that help answer our spoken questions; fraud detection systems that monitor our transactions and alert us if our banking details or identity may be stolen. Whatever flavour of AI and robotics it may be, each is a targeted technology designed to help us. There will be astonishing new AIs and robots to come. They'll be part of our societies, helping us, or even becoming part of us within wearable computing and smart prosthetics.

Not everyone likes this vision of the future. As we saw in chapter 3, some predict that millions of jobs may be lost because of AI and robots. Will factory robots replace factory workers? Will chatbots replace those working in call centres? The answer may well be yes, to these and similar questions. But this process is as old as human technology itself. For every new invention we adopt, we change how people live and work. Old expertise becomes obsolete, new specializations are created. There are few blacksmiths today – there are many factory workers. Tomorrow there may be fewer factory workers and many robot managers and robot maintenance engineers. Today technology changes fast, so our societies

are changing fast. But new opportunities are being generated just as quickly. Driverless cars will need entire new road infrastructures and new ways to sell and maintain them. The computer games industry is booming, with a need for talent in everything from computer simulation to writers and actors. Data storage and analytics – indeed everything and anything to do with data – is thriving, with new kinds of jobs being invented every day. Social media is the new advertising, with new jobs and careers emerging. New forms of interactive entertainment driven by AI and wearable computing are arriving. Every new technology needs new people to create it, set it up, test it, regulate it, use it, fix it. Jobs today are not what they used to be. That's human progress, not AI or robotics.

Ultimately, AI and robotics have always fascinated us and in some cases discomforted us, perhaps because we are interested in ourselves. Uniquely in our technology, building true artificial intelligence may teach us deep truths about how our brains and bodies work. To be successful in building our artificial children we must scrutinize ourselves and how we behave. We will learn how our interactions help us to grow individually and work together in unison. We will learn about emotions, self-awareness and how we make decisions. We will learn about how our actions in the world affect ourselves and others. We will learn what it means to be moral, and how morality and respect for life should be embedded in everything we do. We will learn that we are intimately linked to our environments, which change us, just as we change them. AI is an ongoing voyage of discovery. Perhaps as we travel down this long and difficult path, it might teach us to become better people.

GLOSSARY

Abstraction – the process of simplifying and focusing on only the relevant detail.

Activation function – in artificial neural networks, the output of a neuron is defined by how the activation function transforms the inputs.

Actuator – a component such as an electric motor, hydraulic or pneumatic piston that generates movement.

Affective computing – (also Emotion AI) is the branch of AI dedicated to computer simulation, recognition and processing of emotions.

Agent-based modelling – the use of software agents in combination to simulate a natural or physical system.

Algorithm – the method used by the computer to carry out a specific task.

Artificial immune system – algorithms based on how the immune system works.

Artificial intelligence – (AI), also machine intelligence, intelligent systems, computational intelligence. The study of intelligent agents, or the branches of research in computer science dedicated to enabling computers to achieve what they currently cannot.

Artificial neural network – (ANN) machine-learning algorithms inspired by the workings of neural networks in the brain.

Autonomous car – also self-driving car, robot car or driverless car. An AI-controlled vehicle that may partially or fully take over the task of driving.

Autonomous robot – a robot with behaviours that are performed with a high degree of autonomy.

Backpropagation – the method used in artificial neural networks to train them.

Behaviour tree – a plan that describes how behaviour of a robot or agent is switched between tasks.

Big data – massive data that requires specialized analysis.

Binary tree – a tree data structure in which each node has at most two children.

Boltzmann machine – (also called stochastic Hopfield network with hidden units) is a type of stochastic recurrent neural network that uses symmetrically connected neurons that make random decisions about whether to be on or off.

Capsule neural network – a type of artificial neural network (ANN) that can be used to better model hierarchical relationships, based on biological neural organization.

Chatbot – also smartbots, talkbot, chatterbot, conversational interface or Artificial Conversational Entity. An AI that simulates conversations.

Cognitive science – the interdisciplinary, scientific study of the mind.

Combinatorial optimization – the process of finding one or more optimal solutions in a space of possible solutions.

Combinatorics – the size or enumeration of possible solutions or configurations. Also used in relation to the field of study devoted to enumeration, existence, construction and optimization of arrangements or configurations of structures.

Complex systems – natural or artificial systems that exhibit unpredictable behaviour arising spontaneously from simpler interactions between components.

Computational creativity – also artificial creativity, creative computing. The creation and study of algorithms that produce surprising, unusual or creative results.

Computational neuroscience – also theoretical neuroscience, mathematical neuroscience. The use of mathematical models and theoretical analysis of the brain.

Computer vision – the development of algorithms for processing, interpreting and understanding images and video.

Convolutional neural network – a type of deep neural network algorithm commonly used in vision.

Data science – an interdisciplinary field that aims to extract knowledge and insights from data using mathematics, statistics, information science and computer science.

Datasets – the collection of values associated with features or variables, used by machine learning to learn from. Typically divided into training sets for training the algorithm, validation sets used while tuning parameters of the algorithm, and test sets used to discover the learned accuracy on the data.

Decision-tree learning – an algorithm for predictive modelling using decision trees.

Deep learning – machine-learning methods usually based on ANNs with multiple hidden layers that learn data representations. Learning can be supervised, semi-supervised or unsupervised.

Distributed artificial intelligence – the use of multiple agents working collaboratively to solve a problem. Related to and a predecessor of the field of multi-agent systems.

Effector – also end effector, manipulator, robot hand. The device, often placed at the end of a robot arm, which interacts with the environment. May comprise a gripper to pick up objects or other scientific instruments depending on the function of the robot.

Evolutionary computation – is a family of optimization algorithms inspired by biological evolution, including genetic algorithms, particle-swarm optimization, and ant-colony optimization.

Expert system – a rule-based algorithm designed to resemble the decision-making ability of a human expert by making use of stored knowledge, often in the form of if–then rules.

Fitness function – also evaluation function, objective function. The method for evaluating the quality of a solution, producing a fitness score or value as a result; commonly used in optimization algorithms such as genetic algorithms.

Frames – a data structure used to organize knowledge with 'stereotyped situations'; related to classes in object-oriented programming.

Fuzzy logic – a form of many-valued logic using linguistic variables.

Game theory – is the study of simple models of interaction between rational decision-makers.

Intractable – difficult to solve in practical time. An intractable problem has no efficient algorithm to solve it.

Kernel method – a type of machine-learning algorithm used for pattern analysis, e.g. the support vector machine (SVM).

Knowledge-based system – related to expert systems. A program that reasons, using an inference engine, on a knowledge base, often stored in the form of an ontology or set of rules.

Machine learning – (ML) is the scientific study of algorithms and statistical models to derive or infer patterns or classes within datasets.

Multi-agent system – algorithms making use of multiple interacting intelligent agents that interact to solve a larger problem.

Natural language processing – (NLP), the branch of AI focusing on analysing textual or other unstructured natural language data.

Online machine learning – a type of continuous machine

learning that updates its internal models over time as new data is received, enabling it to adapt to new patterns.

Ontology – a representation, naming and definition of categories, properties and relations between concepts, entities and data.

Reinforcement learning – (RL), a type of machine learning that focuses on how to perform action selection for software agents in an environment in order to maximize a cumulative reward.

Restricted Boltzmann machine – (RBM), is a generative stochastic artificial neural network that can learn a probability distribution over its set of inputs.

Robotics – is an interdisciplinary branch of engineering and science focusing on the design, construction, operation and use of robots, as well as algorithms for their control, sensory feedback and information processing.

Search algorithm – an algorithm that uses the notion of search to find a solution or choice and in doing so optimize or solve a problem.

Semantic web – also data web, web of data. A framework for labelling data within webpages such that it can be shared and reused across applications and systems, and understood by AIs.

Solver – a piece of software that takes problem descriptions as input and calculates their solution, often using search.

Speech recognition – also automatic speech recognition (ASR), computer speech recognition or speech to text (STT). A branch of AI and computational linguistics that enables the recognition and translation of spoken language into text by computers.

Spiking neural network – (SNNs), artificial neural networks designed to be more biologically plausible, by firing chains of spikes at each other.

Strong AI – the hypothesized but currently unrealized form of AI resembling biological intelligence, where general-purpose intelligence is achieved in a computer, and where that intelligence is not simulated.

Symbolic logic – a method of representing logical expressions with symbols and variables, rather than in ordinary language.

Timestep – a discrete chunk of time. Used in discrete models and robot controllers that chop time into manageable chunks to enable the iterative calculation of actions and behaviours in response to sensor values; an example of a timestep duration might be one every millisecond or one every second.

Transistor – an 'electronic switch' that uses an electrical signal to control the flow of electricity. Usually made with semiconductor material such as silicon with at least three connections to a circuit.

Weak AI – also narrow AI. A type of AI where the algorithm is designed to solve a clearly defined task, sometimes by simulating intelligence. All current AI approaches are examples of weak AI.

FURTHER READING

For an easy-to-read history of computer science, including AI:

Digitized: The Science of Computers and How it Shapes Our World, by Peter J. Bentley, Oxford University Press (2012).

For a sensible take on AI today:

Artificial Intelligence: A Guide for Thinking Humans, by Melanie Mitchell, Pelican Books (2019).

For a revealing look at the repercussions of AI:

Rage Inside the Machine: The Prejudice of Algorithms, and How to Stop the Internet Making Bigots of Us All, by Robert Elliot Smith, Bloomsbury Business (2019).

For a realistic view of the limitations of AI today:

Rebooting AI: Building Artificial Intelligence We Can Trust, by Gary Marcus and Ernest Davis, Ballantine Books Inc (2019).

For a fascinating view of the possible future economic and social impact of AI:

AI Superpowers, by Kai-Fu Lee, Houghton Mifflin Harcourt (2018).

For an experienced view of social robots and their use:

A Compassionate Guide for Social Robots, by Marcel Heerink, E3 (2018).

For a view of what it is like to be beaten by an AI, and then to learn from the experience:

Deep Thinking: Where Machine Intelligence Ends and Human Creativity Begins, by Garry Kasparov, John Murray (2017).

For the views of a mathematician on creativity by AI:

The Creativity Code: Art and Innovation in the Age of AI, by Marcus Du Sautoy, Fourth Estate (2019).

For the original and best book on affective computing:

Affective Computing, by Rosalind Picard, MIT Press (2000).

For a classic book on robots and people:

Flesh and Machines: How Robots Will Change Us, by Rodney Brooks, Pantheon Books (2002).

For some detailed explorations inside the Chinese room:

Views into the Chinese Room: New Essays on Searle and Artificial Intelligence, edited by John Preston, Oxford University Press (2002).

For a history of cybernetics and AI:

The Mechanical Mind in History, by Philip Husbands, Owen Holland and Michael Wheeler, MIT Press (2008).

For a classic from the father of AI:

The Society of Mind, by Marvin Minsky, Simon & Schuster (1988).

For a quick but thorough look at machine learning in more depth:

The Hundred-Page Machine Learning Book, by Andriy Burkov (self-published) (2019).

For practical ideas about how chatbots can be used in marketing:

Conversational Marketing: How the World's Fastest Growing Companies Use Chatbots to Generate Leads 24/7/365 (and How You Can Too), by David Cancel and Dave Garhardt, John Wiley and Sons (2019).

For a classic written at the beginning of a new field:
Artificial Life: The Quest for a New Creation, by Steven Levy, Jonathan Cape Ltd (1992).

For a trickier read on causality that has implications for AI:
The Book of Why: The New Science of Cause and Effect, by Judea Pearl and Dana Mackenzie, Penguin Books (2018).

For the musings of a physicist on AI and our future:
Life 3.0: Being Human in the Age of Artificial Intelligence, by Max Tegmark, Penguin Books (2017).

For the view of a nervous philosopher on AI:
Superintelligence: Paths, Dangers, Strategies, by Nick Bostrom, Oxford University Press (2014).

For the view of a sceptical mathematical physicist on AI:
The Emperor's New Mind: Concerning Computers, Minds and the Laws of Physics, by Roger Penrose, Oxford University Press (1999).

For a hugely debated view of the future of AI:
The Singularity is Near, by Raymond Kurzweil, Duckworth (2006).

INDEX

K

L

M

N